Math Refresher
for Adults

The Perfect Solution!

Richard W. Fisher

Edited by Christopher Manhoff

For access to the
Online Video Tutorials,
to go **www.mathessentials.net**
and click on Videos.

Math
Essentials

Math Refresher for Adults

Manufactured in the United States of America

ISBN: 978-0-9994433-6-1

1st printing 2018

Math Essentials
P.O. Box 1723
Los Gatos, CA 95031
Ph 408-314-4573
www.mathessentials.net
math.essentials@verizon.net

Preface

Let's face one big fact…

Math is used in all of our lives. We all shop for clothes, vehicles, and groceries. We all have to plan a budget and learn to spend our money wisely. Determining the best bargain at a store is a mathematical skill that we all have to use on a daily basis. Strong math skills can save us lots of money. Like it or not, we all need to save, spend, and invest our money wisely.

We also need strong math skills in our chosen fields. Whether it be engineering, sales, high tech, medicine, electronics, retail, working in a fast-food-restaurant, or any of thousands of other jobs, they all require essential, foundational math skills.

Worldwide, there is now a huge emphasis on analyzing data when working in a high tech environment. Companies are now relying more and more on data to guide their decisions and trends more than ever. It is necessary for employees to have solid math skills in order to make it possible to analyze this data.

Even though we have calculators and computers, it is essential that employees have a strong set of foundational math skills in order to put these devices to their best use.

Most entry-level jobs require strong basic math skills. Many companies require that potential employees take a math competency test as part of the job application process.

Those who are re-entering the workforce or who are entering college will often have to take a math aptitude or placement test. It is a must to come prepared. Doing well without strong foundational math skills on these tests is nearly impossible.

As far as employers are concerned, being good in math often indicates excellent problem solving skills. All business have problems, and they need to be solved. So, let's face it, math is an important part of our professional lives.

About This Book

This book is designed to ensure that you master the essential, foundational math skills that we all need.

- The book is set up in chapters, with short easy-to-understand lessons. Each lesson is short, concise, and explained in simple language.

- The Table of Contents is detailed to make it easy to find the exact topic that you want to work on.

- Video tutorials are available for each lesson. Just go to **www.mathessentials.net**. Click on the **video lessons button** and use any of the tutorials. The passwords are right on the page, printed in red.

- As a bonus, for more advanced Algebra tutorials, go to **www.nononsensealgebra.com** and you will have access to all videos. **Your access code is D2GH7Y4M3.**

REMEMBER…
This book is a REFRESHER, and is written in a way that will ensure that the most amount of learning will take place in the shortest amount of time. So good luck, and enjoy…

A Few Tips on How to Use This Book Before You Start

- You can be successful in math. This book was written to ensure that you experience that success!

- Use the detailed Table of Contents to help look up a specific topic when necessary

- I recommend to neatly copy problems onto a piece of paper, and then work them. There is something about copying a problem down and then working it that seems to help in accuracy as well as understanding.

- Each lesson has Review Exercises. Some review is necessary to ensure retaining what you have learned.

- Each lesson has two Sample Problems. It is good to work those and check your answers before starting the rest of the lesson.

- Each lesson has a small section to take notes.

- Use the Solutions section in the back of the book to check your answers. A score of at least 80% can be considered a passing score.

- Remember that accuracy is important. But understanding the topic is even more important.

- Remember that there is a corresponding video tutorial for each lesson in this book. For access to the videos, go to **www.mathessentials.net** and click the **videos button**. The **access passwords are shown in red**. For bonus Algebra videos go to **www.nononsensealgebra.com** and **use access code D2GH7Y4M3**. You will have full access to all videos.

- If you are helping a child with a specific topic, viewing and working through the related video together can be quite helpful.

- I recommend, in general, spend no more than an hour or so at a time on math. There is no sense in reaching the point of diminishing returns.

Some Special Tips For Parents Helping Their Kids To Enjoy and Do Well With Math

- Maintain a positive attitude. Help your child you see that math can be fun, and that you enjoy math. Your attitude towards math will have a huge impact on their attitude.

- Link math with daily life. Shopping, cooking, telling time, doing laundry, measuring, making home repairs, and many other tasks, require math skills. Discuss them with your child.

- Make mathematics fun. Math jokes, brain-teasers, and riddles related to math are just a few ideas. Thousands of these can easily be found online.

- Learn about mathematics-related careers. If your child is interested in a specific career, discuss how math relates to that career. Point out how the topic you are working with is related to a specific career. Pilots, physicians, engineers, high tech, and many more professions require excellent math skills.

- Have high expectations for your child. Let your child know that to get good at anything worthwhile, sometimes it involves struggle. Whether it is learning to play a sport or learning to play a musical instrument, at times it can be difficult and that there will be challenges. Life is not always easy.

- Support homework—don't do it! As you work through a problem, ask the child questions, guide them, encourage them, but do not do the work for them.

Contents

— GENERAL MATH —

Percents

Geometry

Integers

Charts and Graphs

Word Problems

— PRE-ALGEBRA and ALGEBRA —

Review Exercises

Notes

1.   362
   + 75

2.   723
   + 12

3.   72
   x   3

4.   6
        4
   +   2

Helpful Hints

1. Line up numbers on the right side.
2. Add the ones first.   * "sum" means to total or add
3. Remember to regroup when necessary.

S. 342	S. 436	1. 42	2. 716

S.   342
      63
   + 512

S.   436
      632
   + 416

1.   42
      53
   + 16

2.   716
      72
   + 314

3.   616
      724
      642
   +   16

4.   723
      436
       19
   +   8

5.   346
      453
      964
   + 234

6.   6
       17
      418
   + 234

7.   362 + 436 + 317 =

8.   27 + 44 + 63 + 73 =

9.   Find the sum of 334, 616, and 743.

10. Find the sum of 16, 19, 23, and 47.

1	
2	
3	
4	
5	
6	
7	
8	
9	
10	
Score	

Problem Solving

There are 33 students in Mr. Jones' class, 41 students in Ms. Martinez's class, and 26 students in Mr. Kelly's class. How many students are there altogether?

Review Exercises

Notes

1.   36
      43
  + 16

2.   367
       19
  + 437

3. 863 + 964 + 17 =

4. Find the sum of 16, 19, 24, and 36

Helpful Hints	When writing large numbers, place commas every three numerals, starting from the right. This makes them easier to read.	**Example:** 21 million, 234 thousand, 416 21,234,416

S.   3,162
       143
  +3,647

S.   7,213
       647
  + 22,134

1. 1,324,167
  + 2,146,179

2. 43,213
     8,137
  +   72

3. 75,643
    3,742
  +76,419

4. 1,736,412
    136,123
  +   7,423

5.   7,163
     14,421
      6,745
  + 14,123

6. 63,742
     1,235
    16,347
  + 2,425

7. Find the sum of 1,342, 176, and 13,417.

8. 17,432 + 7,236 + 6,432 + 16 =

9. 3,672 + 9,876 + 3,712 + 16 =

10. Find the sum of 72,341, 2,342, and 7,963

1	
2	
3	
4	
5	
6	
7	
8	
9	
10	
Score	

Problem Solving	One state has a population of 792,113, and another state has a population of 2,132,415, and a third state has a population of 1,615,500. What is the total population of all three states?

Review Exercises

Notes

1.   3,163
              424
      + 6,734

2. Find the sum of 1,234, 32,164, and 7,321

3.   76,723
         15,342
     + 6,412

4. 3,426 + 73,164 + 17 + 17,650 =

Helpful Hints

1. Line up the numbers on the right side.   * "find the difference" means to subtract
2. Subtract the ones first.
3. Regroup when necessary.   " is how much more" means to subtract

---

S. 643 − 162 S. 3,224 − 763 1. 312 − 71 2. 716 − 317	1
	2
	3
	4
3. 7,163 − 778 4. 7,121 − 1,244 5. 1,493 − 846 6. 4,492 − 1299	5
	6
	7
7. Find the difference between 124 and 86.	8
8. Subtract 426 from 3,496.	9
9. 7,146 − 747 =	10
10. 986 is how much more than 647?	**Score**

Problem Solving

712 students attend Monroe School and 467 students attend Jefferson School. How many more students attend Monroe School than Jefferson School?

Review Exercises

Notes

1.   376
            427
        + 363

2.   716 - 142 =

3. 3,172 + 76 + 7,263 =

4.   6,413
            − 764

Helpful Hints	1. Line up numbers on the right. 2. Subtract the ones first. 3. Sometimes a number must be regrouped more than once.

Examples:

$$\begin{array}{r} {\scriptstyle 6\ 10\ 9\ 1} \\ \cancel{7},\cancel{1}\cancel{0}3 \\ -\ 6\,7\,7 \\ \hline 6\,,4\,2\,6 \end{array} \qquad \begin{array}{r} {\scriptstyle 5\ \ 9\ \ 9} \\ {\scriptstyle \ \ 1\ \ 1\ \ 1\ \ 1} \\ \cancel{6},\cancel{0}\cancel{0}\cancel{0} \\ -1\,,6\,3\,4 \\ \hline 4\,,3\,6\,6 \end{array}$$

S.   701
       − 267

S.   600
       − 379

1.   70
        − 54

2.   603
       − 168

3.   5,013
    − 1,405

4.   500
       − 236

5.   7,000
      − 634

6.   3,102
     −1,634

7. Subtract 7,632 from 12,864.

8. Find the difference between 13,601 and 76,021.

9. 76,102 − 63,234 =

10. What number is 7,001 less than 9,000?

1	
2	
3	
4	
5	
6	
7	
8	
9	
10	
Score	

Problem Solving	A company earned 96,012 dollars in its first year and 123,056 dollars in its second year. How much more did the company earn in its second year than in its first year?

Review Exercises

Notes

1. $701 - 637$

2. $7{,}102 - 673$

3. $343{,}672$
       $72{,}164$
     $+ 736{,}243$

4. $5{,}000 - 467 =$

Helpful Hints   Use what you have learned to solve the following problems.

S. 743
      $7{,}614$
      $16{,}321$
     $+ 5{,}032$

S. $5{,}001 - 1{,}346$

1. 346
      25
     $+ 176$

2. $716 - 143$

1	
2	
3	
4	
5	
6	
7	
8	
9	
10	
Score	

3. $4{,}216$
      764
     $+ 5{,}123$

4. $3{,}732{,}246 + 3{,}510{,}762$

5. $7{,}101 - 1{,}436$

6. Find the difference between 1,964 and 768.

7. $17{,}023 - 13{,}605.$

8.   Find the sum of 236, 742, and 867.

9.   How much more is 763 than 147?

10. $72{,}163 + 16{,}432 + 1{,}963 =$

Problem Solving   A theater has 905 seats. If 693 of them are taken, how many seats are empty?

Review Exercises

Notes

1.   6,032
    − 1,647

2. 364 + 72 + 167 + 396 =

3. Find the difference between 760 and 188.

4.   9,000 - 3,287 =

Helpful Hints	1. Line up numbers on the right. 2. Multiply the ones first. 3. Regroup when necessary. 4. "Product" means to multiply.	**Examples:**

Examples:

$$\begin{array}{r} {}^{1\ 1}644 \\ \times\quad 3 \\ \hline 1,932 \end{array} \qquad \begin{array}{r} {}^{3\ 2}6,076 \\ \times\quad 4 \\ \hline 24,304 \end{array}$$

S.   423   S.   2,345   1.   67   2.   74
    x   3   x   6   x   3   x   6

3.   764   4.   3,142   5.   2,036   6.   3,427
    x   3   x   6   x   8   x   6

7. 8,058 x 7 =

8. 7 x 7,643 =

9. Find the product of 9,746 and 6.

10. Multiply 9 and 13,708.

1	
2	
3	
4	
5	
6	
7	
8	
9	
10	
Score	

Problem Solving	If there are 365 days in each year, then how many days are there in six years?

Review Exercises

Notes

1.   342
   x   6

2.   2034
   x   7

3.   7,103
   − 1,664

4.   32,173
   1,424
   + 3,456

Helpful Hints

1. Line up numbers on the right.
2. Multiply the ones first.
3. Multiply by tens second.
4. Add the two products.

Examples:

```
  43        437
x 32       x 26
  86       2622
1290       8740
1,376     11,362
```

S.   46 x 23   S.   146 x 42   1.   73 x 62   2.   75 x 16

3.   47 x 36   4.   534 x 25   5.   237 x 30   6.   807 x 36

7. Find the product of 16 and 28
8. 38 x 26 =
9. Find the product of 92 and 734
10. 320 x 49 =

1	
2	
3	
4	
5	
6	
7	
8	
9	
10	
Score	

Problem Solving

A school has twenty-six classrooms. If each classroom needs 32 desks, how many desks are needed altogether?

Review Exercises

Notes

1.   23
   x 46

2.   402
   x 36

3.   7,205
   − 1,637

4.   335
        63
       426
   + 173

Helpful Hints

1. Line up numbers on the right.
2. Multiply the ones first.
3. Multiply by tens second.
4. Multiply by hundreds last.
5. Add the products.

Examples:

```
   243
 x 346
  1458
  9720
 72900
84,078
```

```
   673
 x 307
  4711
  0000
201900
206,611
```

S. 132 x 234	S. 623 x 542	1. 233 x 215	2. 326 x 514	**1**	
				2	
				3	
				4	
3. 143 x 203	4. 246 x 403	5. 324 x 616	6. 263 x 300	**5**	
				6	
				7	
7. 361 x 423 =				**8**	
8. Find the product of 306 and 427				**9**	
9. Multiply 600 and 721				**10**	
10. 334 x 466 =				**Score**	

Problem Solving

A factory can produce 215 cars per day. How many cars can it produce in 164 days?

Review Exercises

Notes

1.　　　　306
　　　　　x 7

2.　Find the product of 24
　　and 36

3.　　　7,736
　　　　　493
　　　+2,615

4.　Find the difference
　　between 2,174 and 636

Helpful Hints　　Use what you have learned to solve the following problems.

S.　316 　x 24	S.　604 　x 423	1.　27 　x 6	2.　603 　x 7
3.　3,612 　x 9	4.　63 　x72	5.　263 　x 54	6.　242 　x 643

#	
1	
2	
3	
4	
5	
6	
7	
8	
9	
10	
Score	

7.　Find the product of 12 and 473

8.　600 x 748 =

9.　22 x 410 =

10.　706 x 304 =

Problem Solving　　Each package of paper contains 500 sheets. How many sheets of paper are there in 24 packages?

Review Exercises

Notes

1.   712
       − 463

2.   320
       x 6

3.   65,426
     + 73,437

4.   Find the product of 26 and 37

Helpful Hints

1. Divide.
2. Multiply.
3. Subtract.
4. Begin again.

Examples:

$$\begin{array}{r} 15\,r2 \\ 3\overline{)47} \\ -3\downarrow \\ \hline 17 \\ -15 \\ \hline 2 \end{array}$$

$$\begin{array}{r} 9\,r5 \\ 6\overline{)59} \\ -54 \\ \hline 5 \end{array}$$

REMEMBER! The remainder must be less than the divisor.

S. $3\overline{)16}$ S. $7\overline{)69}$ 1. $4\overline{)34}$ 2. $8\overline{)43}$

3. $7\overline{)87}$ 4. $6\overline{)93}$ 5. $8\overline{)97}$ 6. $6\overline{)43}$

7. $66 \div 7 =$

8. $97 \div 4 =$

9. $\dfrac{61}{5} =$

10. $\dfrac{37}{2} =$

1	
2	
3	
4	
5	
6	
7	
8	
9	
10	
Score	

Problem Solving

A teacher needs 72 rulers for his class. If rulers come in boxes that contain six rulers, how many boxes of rulers does the teacher need?

Review Exercises

Notes

1. $6\overline{)39}$　　　　　2. $5\overline{)79}$

3. $7 \times 2{,}134 =$　　　　　4. $6{,}000 - 768 =$

Helpful Hints

1. Divide.
2. Multiply.
3. Subtract.
4. Begin again.

Examples:

$$\begin{array}{r} 171\,r2 \\ 3\overline{)515} \\ -3\!\downarrow\!\downarrow \\ \hline 21 \\ -21 \\ \hline 05 \\ -3 \\ \hline 2 \end{array}$$

$$\begin{array}{r} 203 \\ 4\overline{)812} \\ -8\!\downarrow\!\downarrow \\ \hline 01 \\ -0 \\ \hline 12 \\ -12 \\ \hline 0 \end{array}$$

REMEMBER! The remainder must be less than the divisor.

S. $3\overline{)432}$　S. $6\overline{)237}$　1. $3\overline{)952}$　2. $2\overline{)819}$

3. $7\overline{)924}$　4. $6\overline{)817}$　5. $5\overline{)727}$　6. $6\overline{)950}$

7. $4\overline{)484}$　8. $9\overline{)886}$　9. $8\overline{)979}$　10. $6\overline{)953}$

1	
2	
3	
4	
5	
6	
7	
8	
9	
10	
Score	

Problem Solving

A theater has 325 seats They are placed in 9 equal rows. How many seats are in each row? How many seats will be left over?

Review Exercises

Notes

1. $3\overline{)79}$ 2. $6\overline{)557}$

3. $\begin{array}{r} 46 \\ \times\ 23 \\ \hline \end{array}$

4. Find the difference between 236 and 84

Examples:

1. Divide.
2. Multiply.
3. Subtract.
4. Begin again.

$$\begin{array}{r} 1708 \\ 3\overline{)5124} \\ -3\downarrow\downarrow\downarrow \\ \hline 21 \\ -21 \\ \hline 02 \\ -0 \\ \hline 24 \\ -24 \\ \hline 0 \end{array}$$

$$\begin{array}{r} 448\,r2 \\ 4\overline{)1794} \\ -16\downarrow\downarrow \\ \hline 19 \\ -16 \\ \hline 34 \\ -32 \\ \hline 2 \end{array}$$

REMEMBER! The remainder must be less than the divisor.

Helpful Hints

S. $3\overline{)7062}$ S. $4\overline{)3452}$ 1. $2\overline{)1132}$ 2. $2\overline{)6743}$

3. $7\overline{)7854}$ 4. $6\overline{)2319}$ 5. $5\overline{)6555}$ 6. $4\overline{)5995}$

7. $7\overline{)73,172}$ 8. $2\overline{)23,960}$ 9. $7\overline{)71,345}$ 10. $6\overline{)32,106}$

1	
2	
3	
4	
5	
6	
7	
8	
9	
10	
Score	

Problem Solving

Mrs. Arnold made 1,296 cookies. If she puts them into boxes that contain 9 cookies each, how many boxes will she need?

Review Exercises

Notes

1. $7\overline{)1422}$

2. $\begin{array}{r} 206 \\ \times\ 36 \\ \hline \end{array}$

3. $\begin{array}{r} 710 \\ -\ 167 \\ \hline \end{array}$

4. $3{,}752 + 17{,}343 + 964 =$

Helpful Hints

1. Divide.
2. Multiply.
3. Subtract.
4. Begin again.

* Remainders must be less than the divisor
* Zeroes may sometimes appear in the quotient

S. $3\overline{)245}$ S. $8\overline{)8568}$ 1. $6\overline{)302}$ 2. $5\overline{)7500}$

3. $3\overline{)6314}$ 4. $9\overline{)3767}$ 5. $7\overline{)1563}$ 6. $8\overline{)716}$

7. $6\overline{)1817}$ 8. $6\overline{)4793}$ 9. $6\overline{)6007}$ 10. $8\overline{)3207}$

1	
2	
3	
4	
5	
6	
7	
8	
9	
10	
Score	

Problem Solving

Six people in a club each sold the same number of tickets. If 636 tickets were sold, how many tickets did each person sell?

Review Exercises

1. $7\overline{)818}$ 2. $2{,}003 - 765 =$

3. $\begin{array}{r}453\\ \times\ 600\\\hline\end{array}$ 4. $4\overline{)4003}$

Helpful Hints

1. Divide.
2. Multiply.
3. Subtract.
4. Begin again.

Examples:

$$\begin{array}{r}12\ \text{r}49\\60\overline{)769}\\-\ 60\downarrow\\\hline169\\-\ 120\\\hline49\end{array}\qquad\begin{array}{r}44\ \text{r}5\\40\overline{)1765}\\-\ 160\downarrow\\\hline165\\-\ 160\\\hline5\end{array}$$

S. $30\overline{)187}$ S. $40\overline{)5342}$ 1. $70\overline{)342}$ 2. $60\overline{)399}$

3. $50\overline{)463}$ 4. $50\overline{)727}$ 5. $30\overline{)1763}$ 6. $20\overline{)1423}$

7. $50\overline{)8324}$ 8. $90\overline{)9281}$ 9. $50\overline{)4751}$ 10. $50\overline{)2526}$

1	
2	
3	
4	
5	
6	
7	
8	
9	
10	
Score	

Problem Solving

A bakery put cookies into boxes of 20 each. If 1,720 cookies were baked, how many boxes would be needed?

Review Exercises

Notes

1. $\begin{array}{r} 643 \\ 76 \\ + 492 \\ \hline \end{array}$

2. $\begin{array}{r} 403 \\ - 247 \\ \hline \end{array}$

3. $\begin{array}{r} 675 \\ \times 32 \\ \hline \end{array}$

4. $7\overline{)8172}$

Helpful Hints

Sometimes it is easier to mentally round the divisor to the nearest multiple of ten.

Example:

$$\begin{array}{r} 32\,r11 \\ 22\overline{)715} \\ -66\downarrow \\ \hline 55 \\ -44 \\ \hline 11 \end{array}$$

Think of

$20\overline{)715}$

S. $31\overline{)672}$ S. $22\overline{)684}$ 1. $41\overline{)927}$ 2. $28\overline{)603}$

3. $68\overline{)743}$ 4. $93\overline{)692}$ 5. $43\overline{)913}$ 6. $31\overline{)984}$

7. $62\overline{)724}$ 8. $43\overline{)501}$ 9. $21\overline{)946}$ 10. $41\overline{)866}$

1	
2	
3	
4	
5	
6	
7	
8	
9	
10	
Score	

Problem Solving

There are 416 students in a school. If there are 32 students in each class, then how many classes are there?

Review Exercises

Notes

1. $2\overline{)716}$ 2. $30\overline{)524}$

3. $31\overline{)671}$ 4. $39\overline{)791}$

Helpful Hints

Sometimes it is necessary to correct your estimate.

Example:

$$63\overline{)\begin{array}{r} 6 \\ 374 \\ -378 \end{array}}\text{ -- Too Large}$$

$$63\overline{)\begin{array}{r} 5\ ^{r60} \\ 375 \\ -315 \\ \hline 60 \end{array}}$$

S. $74\overline{)293}$ S. $43\overline{)821}$ 1. $38\overline{)197}$ 2. $87\overline{)522}$

1	
2	
3	
4	
5	
6	
7	
8	
9	
10	
Score	

3. $18\overline{)997}$ 4. $21\overline{)178}$ 5. $32\overline{)163}$ 6. $14\overline{)886}$

7. $34\overline{)649}$ 8. $42\overline{)829}$ 9. $36\overline{)721}$ 10. $18\overline{)787}$

Problem Solving

If each carton contains 36 eggs, how many eggs are there in twenty-four cartons?

Review Exercises

Notes

1. $30\overline{)96}$ 2. $40\overline{)542}$

3. $40\overline{)396}$ 4. $57\overline{)361}$

Use what you have learned to solve the following problems. Remember to mentally round your divisor to the nearest multiple of ten.

Example:

$$
\begin{array}{r}
216\ \text{r}20 \\
32\overline{)6932} \\
-64\downarrow\downarrow \\
\hline
53 \\
-32 \\
\hline
212 \\
-192 \\
\hline
20
\end{array}
$$

Think of

$30\overline{)6932}$

Helpful Hints

S. $44\overline{)9350}$ S. $31\overline{)6432}$ 1. $23\overline{)2645}$ 2. $41\overline{)9597}$	1
	2
	3
	4
3. $48\overline{)4896}$ 4. $31\overline{)4133}$ 5. $27\overline{)8191}$ 6. $18\overline{)5508}$	5
	6
	7
7. $21\overline{)1537}$ 8. $32\overline{)1286}$ 9. $49\overline{)2222}$ 10. $25\overline{)1049}$	8
	9
	10
	Score

Problem Solving

Sixteen bales of hay weigh a total of 2,000 pounds. How much does each bale weigh if they are all the same size?

Review Exercises

Notes

1. 3$\overline{)72}$

2. 6$\overline{)1792}$

3. 40$\overline{)5436}$

4. 21$\overline{)899}$

Helpful Hints	Use what you have learned to solve the following problems.	**Example:**	* Mentally round 2-digit divisors to the nearest multiple of ten. * Remainders must be less than the divisor.

S. 81$\overline{)7816}$ S. 28$\overline{)1661}$ 1. 2$\overline{)87}$ 2. 7$\overline{)1936}$

3. 5$\overline{)5127}$ 4. 30$\overline{)547}$ 5. 70$\overline{)196}$ 6. 40$\overline{)3379}$

7. 38$\overline{)9699}$ 8. 23$\overline{)2633}$ 9. 61$\overline{)2,222}$ 10. 32$\overline{)7049}$

1	
2	
3	
4	
5	
6	
7	
8	
9	
10	
Score	

Problem Solving	A car can travel 32 miles using one gallon of gasoline. How many gallons of gasoline will be needed to travel 512 miles?

1. $\begin{array}{r} 342 \\ 53 \\ + \ 616 \\ \hline \end{array}$ 2. $\begin{array}{r} 746 \\ 716 \\ 823 \\ + \ 634 \\ \hline \end{array}$ 3. $7{,}362 + 775 + 72{,}516 =$

4. $7{,}013 + 2{,}615 + 776 + 29 =$ 5. $7{,}001 + 696 + 18 + 732 =$

6. $\begin{array}{r} 743 \\ - \ 367 \\ \hline \end{array}$ 7. $\begin{array}{r} 5{,}282 \\ - \ 1{,}367 \\ \hline \end{array}$ 8. $7{,}052 - 2{,}637 =$

9. $6{,}000 - 3{,}678 =$ 10. $7{,}001 - 678 =$

11. $\begin{array}{r} 76 \\ \times \ 3 \\ \hline \end{array}$ 12. $\begin{array}{r} 7{,}653 \\ \times \ 4 \\ \hline \end{array}$ 13. $\begin{array}{r} 53 \\ \times \ 46 \\ \hline \end{array}$ 14. $\begin{array}{r} 627 \\ \times \ 36 \\ \hline \end{array}$

15. $\begin{array}{r} 673 \\ \times \ 346 \\ \hline \end{array}$ 16. $3\overline{)425}$ 17. $6\overline{)1697}$ 18. $30\overline{)769}$

19. $42\overline{)8992}$ 20. $28\overline{)1577}$

1	
2	
3	
4	
5	
6	
7	
8	
9	
10	
11	
12	
13	
14	
15	
16	
17	
18	
19	
20	

Review Exercises

Notes

1.   701
      − 267

2.   337
        756
      +   63

3.   3) 4162

4.   48
      x   27

Helpful Hints	A fraction is a number that names a part of a whole or a group	**Example:**  $= \dfrac{3}{4} \begin{matrix}\leftarrow \text{numerator} \\ \leftarrow \text{denominator}\end{matrix}$ *Think of $\dfrac{3}{4}$ as $\dfrac{3 \text{ of}}{4 \text{ equal parts}}$

Write a fraction for each shaded figure (some may have more than one name).

1	
2	
3	
4	
5	
6	
7	
8	
9	
10	
Score	

S.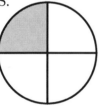

S.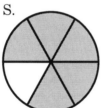

1.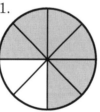

2.

3.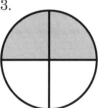

4.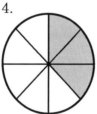

5.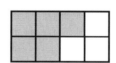

6.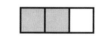

7.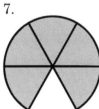

8.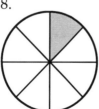

9.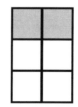

10.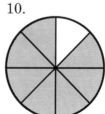

Problem Solving

If a box of crayons holds 24 crayons, how many crayons are there in sixteen boxes?

Review Exercises

Notes

1.   7 x 306 =

2.   72 + 316 + 726 =

3.   810 − 316 =

4. $20 \overline{)317}$

Helpful Hints

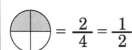 $= \frac{2}{4} = \frac{1}{2}$

$\frac{2}{4}$ has been reduced to its simplest form which is $\frac{1}{2}$. Divide the numerator and denominator by the largest possible number.

Examples: $2 \overline{)\frac{6}{8}} = \frac{3}{4}$

Sometimes more than one step can be used:

$2 \overline{)\frac{24}{28}} = 2 \overline{)\frac{12}{14}} = \frac{6}{7}$

Reduce each fraction to its lowest terms.

S. $\frac{5}{10} =$

S. $\frac{12}{16} =$

1. $\frac{12}{15} =$

2. $\frac{15}{20} =$

3. $\frac{10}{20} =$

4. $\frac{20}{25} =$

5. $\frac{12}{18} =$

6. $\frac{16}{24} =$

7. $\frac{24}{40} =$

8. $\frac{20}{32} =$

9. $\frac{15}{18} =$

10. $\frac{18}{24} =$

1	
2	
3	
4	
5	
6	
7	
8	
9	
10	
Score	

Problem Solving

If there are 24 crayons in each box, how many crayons are there in $2\frac{1}{2}$ boxes?

Review Exercises

Notes

1. What fraction of the figure is shaded?

2. Reduce $\frac{6}{9}$ to its lowest terms.

3. $32 \overline{)\ 679}$

4.
$$\begin{array}{r} 216 \\ \times\ 304 \\ \hline \end{array}$$

Helpful Hints

An improper fraction has a numerator that is greater than or equal to its denominator. An improper fraction can be written either as a whole number or as a mixed numeral (a whole number and a fraction).

Example: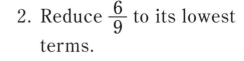

*Divide the numerator by the denominator

Change each fraction to a mixed numeral or whole number.

S. $\frac{7}{4} =$ S. $\frac{9}{6} =$ 1. $\frac{10}{4} =$ 2. $\frac{10}{7} =$

3. $\frac{30}{15} =$ 4. $\frac{24}{5} =$ 5. $\frac{18}{4} =$ 6. $\frac{36}{12} =$

7. $\frac{45}{10} =$ 8. $\frac{38}{6} =$ 9. $\frac{12}{8} =$ 10. $\frac{60}{25} =$

1	
2	
3	
4	
5	
6	
7	
8	
9	
10	
Score	

Problem Solving

On Saturday, 20,136 people visited the zoo. On Sunday, there were 17,308 visitors. How many more people visited the zoo on Saturday than on Sunday.

Review Exercises

Notes

1. Change $\frac{9}{7}$ to a mixed numeral

2. Change $\frac{25}{20}$ to a mixed numeral

3. Reduce $\frac{25}{35}$ to its lowest terms

4. What fraction of the figure is shaded?

Helpful Hints	To add fractions with like denominators, add the numerators and then ask the following questions: 1. Is the fraction improper? If it is, make it a mixed numeral or whole number. 2. Can the fraction be reduced? If it can, reduce it to its simplest form.	**Example:** $\frac{7}{10}$ $+\frac{5}{10}$ $\frac{12}{10} = 1\frac{2}{10} = 1\frac{1}{5}$

S. $\frac{1}{10}$ $+\frac{5}{10}$

S. $\frac{3}{7}$ $+\frac{5}{7}$

1. $\frac{7}{12}$ $+\frac{2}{12}$

2. $\frac{3}{8} + \frac{3}{8}$

3. $\frac{2}{5}$ $+\frac{1}{5}$

4. $\frac{5}{12}$ $+\frac{1}{12}$

5. $\frac{5}{8}$ $+\frac{5}{8}$

6. $\frac{11}{16}$ $+\frac{13}{16}$

7. $\frac{1}{3}$ $\frac{2}{3}$ $+\frac{2}{3}$

8. $\frac{1}{4}$ $\frac{3}{4}$ $+\frac{3}{4}$

9. $\frac{7}{8}$ $+\frac{7}{8}$

10. $\frac{5}{6}$ $+\frac{5}{6}$

1	
2	
3	
4	
5	
6	
7	
8	
9	
10	
Score	

Problem Solving

At Lincoln school $\frac{1}{9}$ of the students ride their bikes to school, $\frac{5}{9}$ of the students walk or ride their bikes to school. What faction of the students either walk or ride their bikes to school? (Be sure your answer is expressed in its simplest form.)

Review Exercises

Notes

1. $\dfrac{2}{5}$
    $+\dfrac{1}{5}$
    _____

2. $60\,\overline{\smash{)}\,729}$

3. $\dfrac{7}{10}$
    $+\dfrac{5}{10}$
    _____

4. $7 + 19 + 342 =$

Helpful Hints

1. Add the fractions first.
2. Add the whole numbers next.
3. It there is an improper fraction, change it to a mixed numeral.
4. Add the mixed numeral to the whole number.

Example: $6\frac{7}{8}$
$+2\frac{5}{8}$
$8\frac{12}{8} = 8 + 1\frac{4}{8} = 9\frac{4}{8} = 9\frac{1}{2}$

*Reduce fractions to lowest terms

1	
2	
3	
4	
5	
6	
7	
8	
9	
10	
Score	

S. $3\frac{1}{4}$
$+\,2\frac{1}{4}$
_____

S. $3\frac{5}{8}$
$+\,2\frac{3}{8}$
_____

1. $3\frac{2}{5}$
$+\,4\frac{3}{5}$
_____

2. $4\frac{3}{4}$
$+\,2\frac{3}{4}$
_____

3. $3\frac{7}{10}$
$+\,4\frac{3}{10}$
_____

4. $5\frac{5}{6}$
$+\,2\frac{2}{6}$
_____

5. $6\frac{7}{9}$
$+\,5\frac{5}{9}$
_____

6. $3\frac{4}{5}$
$+\,2\frac{2}{5}$
_____

7. $5\frac{9}{10}$
$+\,3\frac{3}{10}$
_____

8. $7\frac{5}{6}$
$+\,4\frac{4}{6}$
_____

9. $7\frac{1}{2}$
$+\,2\frac{1}{2}$
_____

10. $3\frac{5}{7}$
$+\,4\frac{3}{7}$
_____

Problem Solving

A baker uses $\frac{7}{8}$ cups of flour for each pie he bakes. He uses $\frac{3}{8}$ cups of flour for each cake. How much flour is used to make 2 pies and 1 cake?

Review Exercises

Notes

1. Reduce $\frac{15}{18}$ to its lowest terms

2. Change $\frac{19}{5}$ to a mixed numeral

3. $\begin{array}{r} \frac{3}{4} \\ + \frac{3}{4} \\ \hline \end{array}$

4. $\begin{array}{r} 2\frac{3}{5} \\ + 4\frac{3}{5} \\ \hline \end{array}$

Helpful Hints

To subtract fractions that have like denominators, subtract the numerators. Reduce the fractions to their lowest terms.

Example: $\begin{array}{r} \frac{9}{10} \\ - \frac{3}{10} \\ \hline \frac{6}{10} = \frac{3}{5} \end{array}$

S. $\begin{array}{r} \frac{7}{8} \\ - \frac{3}{8} \\ \hline \end{array}$

S. $\begin{array}{r} \frac{3}{4} \\ - \frac{1}{4} \\ \hline \end{array}$

1. $\begin{array}{r} \frac{5}{8} \\ - \frac{1}{8} \\ \hline \end{array}$

2. $\begin{array}{r} \frac{9}{10} \\ - \frac{1}{10} \\ \hline \end{array}$

3. $\begin{array}{r} \frac{7}{8} \\ - \frac{1}{8} \\ \hline \end{array}$

4. $\frac{8}{9} - \frac{2}{9} =$

5. $\begin{array}{r} \frac{19}{20} \\ - \frac{4}{20} \\ \hline \end{array}$

6. $\begin{array}{r} \frac{7}{11} \\ - \frac{3}{11} \\ \hline \end{array}$

7. $\begin{array}{r} \frac{11}{12} \\ - \frac{5}{12} \\ \hline \end{array}$

8. $\begin{array}{r} \frac{6}{7} \\ - \frac{1}{7} \\ \hline \end{array}$

9. $\begin{array}{r} \frac{23}{24} \\ - \frac{13}{24} \\ \hline \end{array}$

10. $\begin{array}{r} \frac{11}{16} \\ - \frac{7}{16} \\ \hline \end{array}$

1	
2	
3	
4	
5	
6	
7	
8	
9	
10	
Score	

Problem Solving

John walks $\frac{9}{10}$ miles to school. If he has already gone $\frac{3}{10}$ miles, how much farther does he have to walk before he gets to school?

35

Review Exercises

Notes

1. $\frac{3}{4}$
    $-\frac{1}{4}$
    ___

2. $\frac{3}{5}$
    $+\frac{3}{5}$
    ___

3. $2\frac{3}{8}$
    $+3\frac{1}{8}$
    ___

4. $1{,}708 - 765 =$

Helpful Hints	To subtract numerals with like denominators, subtract the fractions first, then the whole numbers. Reduce fractions to lowest terms. If the fractions can't be subtracted, take one from the whole number, increase the fraction, then subtract.	**Examples:** $7\frac{3}{4}$ $-2\frac{1}{4}$ $5\frac{2}{4} = 5\frac{1}{2}$	$6\cancel{7}\frac{1}{4} + \frac{4}{4} = \frac{5}{4}$ $-2\frac{3}{4}$ $4\frac{2}{4} = 4\frac{1}{2}$

S. $3\frac{3}{4}$
   $-1\frac{1}{4}$
   ___

S. $5\frac{1}{3}$
   $-2\frac{2}{3}$
   ___

1. $7\frac{5}{8}$
   $-1\frac{3}{8}$
   ___

2. $6\frac{1}{4}$
   $-1\frac{3}{4}$
   ___

3. $7\frac{7}{15}$
   $-2\frac{11}{15}$
   ___

4. $8\frac{8}{9}$
   $-2\frac{2}{9}$
   ___

5. $6\frac{4}{5}$
   $-2\frac{2}{5}$
   ___

6. $7\frac{1}{10}$
   $-3\frac{7}{10}$
   ___

7. $4\frac{1}{7}$
   $-2\frac{1}{7}$
   ___

8. $7\frac{3}{10}$
   $-4\frac{7}{10}$
   ___

9. $6\frac{1}{5}$
   $-2\frac{3}{5}$
   ___

10. $7\frac{7}{15}$
    $-4\frac{13}{15}$
    ___

1	
2	
3	
4	
5	
6	
7	
8	
9	
10	
Score	

Problem Solving

A woman worked $2\frac{2}{3}$ hours on Monday and $3\frac{2}{3}$ hours on Tuesday. How many hours did she work altogether?

Review Exercises

Notes

1. $\dfrac{1}{4}$

    $+\ \dfrac{1}{4}$
    _____

2. $\dfrac{3}{8}$

    $+\ \dfrac{7}{8}$
    _____

3. $4\dfrac{2}{3}$

    $+\ 3\dfrac{2}{3}$
    _____

4. $6\dfrac{3}{4}$

    $+\ 5\dfrac{3}{4}$
    _____

Helpful Hints

To subtract a fraction or mixed numeral from a whole number, take one from the whole number and make it a fraction, then subtract.

Examples:

$$\overset{3}{\cancel{4}} \to \dfrac{4}{4}$$
$$-\ 2\quad\dfrac{1}{4}$$
$$\overline{1\dfrac{3}{4}}$$

$$\overset{6}{\cancel{7}} \to \dfrac{5}{5}$$
$$-\quad\dfrac{3}{5}$$
$$\overline{6\dfrac{2}{5}}$$

S. 6

$-\ 2\dfrac{3}{5}$
_____

S. 7

$-\ \dfrac{3}{4}$
_____

1. 6

$-\ 2\dfrac{4}{7}$
_____

2. 5

$-\ 1\dfrac{3}{5}$
_____

3. 7

$-\ \dfrac{2}{3}$
_____

4. 6

$-\ 2\dfrac{9}{10}$
_____

5. 7

$-\ 2\dfrac{1}{8}$
_____

6. 16

$-\ 12\dfrac{3}{8}$
_____

7. 7

$-\ 3\dfrac{7}{9}$
_____

8. 4

$-\ 3\dfrac{1}{2}$
_____

9. 6

$-\ 2\dfrac{3}{10}$
_____

10. 5

$-\ \dfrac{3}{5}$
_____

1	
2	
3	
4	
5	
6	
7	
8	
9	
10	
Score	

Problem Solving

A tailor had 7 yards of cloth. He used $4\dfrac{7}{8}$ yards to make a suit. How many yards were left?

Review Exercises

Notes

1.
$$\frac{7}{8}$$
$$-\frac{2}{8}$$
___

2.
$$\frac{3}{4}$$
$$+\frac{1}{4}$$
___

3. Convert $\frac{14}{10}$ to a mixed numeral.

4. Reduce $\frac{16}{20}$ to its lowest terms.

Helpful Hints

Use what you have learned to solve the following problems. Regroup when necessary.

Reduce all answers to their lowest terms. If problems are positioned horizontally, put them in columns before working.

S. $2\frac{7}{10}$ $+5\frac{5}{10}$

S. $7\frac{1}{3}$ $-2\frac{2}{3}$

1. $\frac{7}{9}$ $+\frac{3}{9}$

2. $\frac{15}{16}$ $-\frac{3}{16}$

3. $3\frac{3}{5}$ $+7\frac{3}{5}$

4. 7 $-2\frac{1}{3}$

5. $6\frac{5}{8}$ $-1\frac{1}{8}$

6. $5-\frac{1}{3} =$

7. $3\frac{7}{8}$ $+4\frac{5}{8}$

8. $6\frac{1}{3}$ $-1\frac{2}{3}$

9. $7\frac{1}{10}$ $-3\frac{7}{10}$

10. $\frac{3}{5}+\frac{4}{5}+\frac{3}{5} =$

1	
2	
3	
4	
5	
6	
7	
8	
9	
10	
Score	

Problem Solving

A family has $12\frac{1}{3}$ pounds of beef in the freezer. If they used $3\frac{2}{3}$ pounds for dinner, how much beef is left?

Review Exercises

Notes

1. Find the sum of $\dfrac{4}{5}$ and $\dfrac{3}{5}$.

2. Find the difference between $\dfrac{7}{8}$ and $\dfrac{3}{8}$.

3.
$$\begin{array}{r} 7 \\ -\ 2\ \dfrac{3}{4} \\ \hline \end{array}$$

4.
$$\begin{array}{r} 7\ \dfrac{3}{5} \\ -\ 4 \\ \hline \end{array}$$

Helpful Hints

To add or subtract fractions with unlike denominators you need to find the least common denominator (LCD). The LCD is the smallest number, other than zero, that each denominator will divide into evenly.

Examples:
The Least Common Denominator of:
$\frac{1}{5}$ and $\frac{1}{10}$ is 10 $\frac{3}{8}$ and $\frac{1}{6}$ is 24

Find the least common denominator of each of the following:

S. $\dfrac{1}{3}$ and $\dfrac{3}{4}$ S. $\dfrac{3}{8}$ and $\dfrac{7}{12}$ 1. $\dfrac{1}{5}$ and $\dfrac{4}{15}$ 2. $\dfrac{5}{6}$ and $\dfrac{7}{9}$

3. $\dfrac{9}{14}$ and $\dfrac{1}{7}$ 4. $\dfrac{1}{8}, \dfrac{1}{6}$ and $\dfrac{1}{4}$ 5. $\dfrac{5}{9}, \dfrac{5}{6}$ and $\dfrac{7}{12}$ 6. $\dfrac{4}{5}$ and $\dfrac{1}{4}$

7. $\dfrac{1}{13}$ and $\dfrac{7}{39}$ 8. $\dfrac{5}{12}, \dfrac{7}{20}$ and $\dfrac{11}{60}$ 9. $\dfrac{11}{24}, \dfrac{3}{16}$ and $\dfrac{13}{48}$ 10. $\dfrac{1}{9}, \dfrac{5}{12}$ and $\dfrac{5}{6}$

1	
2	
3	
4	
5	
6	
7	
8	
9	
10	
Score	

Problem Solving

A plane traveled 4,500 miles in six hours. What was its average speed per hour?

Review Exercises

Notes

1. $3\overline{)602}$

2. $\begin{array}{r} 43 \\ \times\ 36 \\ \hline \end{array}$

3. $36 + 7 + 309 =$

4. $600 - 139 =$

Helpful Hints

To add or subtract fractions with unlike denominators, find the least common denominator. Multiply each fraction by one to make equivalent fractions. Finally, add or subtract.

Examples:

$$\frac{2}{5} \times \frac{2}{2} = \frac{4}{10}$$
$$+\frac{1}{2} \times \frac{5}{5} = \frac{5}{10}$$
$$\overline{\qquad} \qquad \frac{9}{10}$$

$$\frac{5}{6} \times \frac{2}{2} = \frac{10}{12}$$
$$+\frac{1}{4} \times \frac{3}{3} = \frac{3}{12}$$
$$\overline{\qquad} \qquad \frac{13}{12} = 1\frac{1}{12}$$

S. $\begin{array}{r} \frac{1}{3} \\ +\ \frac{1}{4} \\ \hline \end{array}$

S. $\begin{array}{r} \frac{4}{5} \\ -\ \frac{3}{10} \\ \hline \end{array}$

1. $\begin{array}{r} \frac{2}{9} \\ +\ \frac{1}{3} \\ \hline \end{array}$

2. $\begin{array}{r} \frac{2}{3} \\ -\ \frac{1}{2} \\ \hline \end{array}$

3. $\begin{array}{r} \frac{5}{6} \\ +\ \frac{1}{3} \\ \hline \end{array}$

4. $\begin{array}{r} \frac{2}{5} \\ +\ \frac{1}{3} \\ \hline \end{array}$

5. $\begin{array}{r} \frac{5}{6} \\ -\ \frac{5}{12} \\ \hline \end{array}$

6. $\begin{array}{r} \frac{1}{2} \\ +\ \frac{4}{7} \\ \hline \end{array}$

7. $\begin{array}{r} \frac{4}{5} \\ +\ \frac{7}{10} \\ \hline \end{array}$

8. $\begin{array}{r} \frac{3}{11} \\ +\ \frac{1}{2} \\ \hline \end{array}$

9. $\begin{array}{r} \frac{4}{7} \\ -\ \frac{1}{2} \\ \hline \end{array}$

10. $\begin{array}{r} \frac{8}{9} \\ +\ \frac{1}{4} \\ \hline \end{array}$

1	
2	
3	
4	
5	
6	
7	
8	
9	
10	
Score	

Problem Solving

John bought nine gallons of paint to paint his house. If he used $5\frac{3}{8}$ gallons, how much does he have left?

Review Exercises

Notes

1. $20\overline{)3762}$

2. $\dfrac{7}{8}$
      $-\dfrac{1}{8}$
      _____

3. $\dfrac{9}{10}$
      $+\dfrac{1}{5}$
      _____

4. $\dfrac{3}{4}$
      $-\dfrac{1}{3}$
      _____

Helpful Hints

When adding mixed numerals with unlike denominators, first add the fractions. If there is an improper fraction, make it a mixed numeral. Finally, add the sum to the sum of the whole numbers.
*Reduce fractions to lowest terms.

Example:

$3\dfrac{2}{3} \times \dfrac{2}{2} = \dfrac{4}{6}$

$+2\dfrac{1}{2} \times \dfrac{3}{3} = \dfrac{3}{6}$

$5 \qquad \dfrac{7}{6} = 1\dfrac{1}{6} = 6\dfrac{1}{6}$

S. $3\dfrac{2}{3}$
    $+4\dfrac{1}{4}$
    _____

S. $4\dfrac{1}{2}$
    $+3\dfrac{3}{5}$
    _____

1. $5\dfrac{5}{6}$
    $+2\dfrac{1}{3}$
    _____

2. $7\dfrac{1}{4}$
    $+3\dfrac{1}{2}$
    _____

3. $5\dfrac{7}{8}$
    $+2\dfrac{1}{4}$
    _____

4. $6\dfrac{3}{7}$
    $+2\dfrac{1}{14}$
    _____

5. $8\dfrac{1}{4}$
    $+7\dfrac{1}{2}$
    _____

6. $7\dfrac{3}{10}$
    $+2\dfrac{7}{20}$
    _____

7. $3\dfrac{1}{5}$
    $+2\dfrac{1}{10}$
    _____

8. $7\dfrac{7}{9}$
    $+3\dfrac{5}{18}$
    _____

9. $6\dfrac{1}{3}$
    $+2\dfrac{1}{5}$
    _____

10. $9\dfrac{3}{4}$
    $+2\dfrac{1}{6}$
    _____

1	
2	
3	
4	
5	
6	
7	
8	
9	
10	
Score	

Problem Solving

A factory can produce 352 parts each hour. How many parts can it produce in 12 hours?

Review Exercises

Notes

1. $\dfrac{3}{7}$

   $+\dfrac{1}{2}$
   —————

2. $5\dfrac{7}{8}$

   $+\ 4\dfrac{1}{8}$
   —————

3. 3

   $+1\dfrac{2}{5}$
   —————

4. $4\dfrac{1}{3}$

   $-2\dfrac{2}{3}$
   —————

Helpful Hints

To subtract mixed numerals with unline denominators, first subtract the fractions. If the fractions cannot be subtracted, take one from the whole number, increase the fraction, then subtract.

Examples:

$$\overset{5}{\cancel{6}}\dfrac{1}{6} = \dfrac{2}{12} + \dfrac{12}{12} = \dfrac{14}{12}$$
$$-\ 3\dfrac{1}{4} = \dfrac{3}{12}$$
$$\overline{\quad 2\dfrac{11}{12}}$$

$$7\dfrac{1}{2} \times \dfrac{3}{3} = \dfrac{3}{6}$$
$$-\ 2\dfrac{1}{3} \times \dfrac{2}{2} = \dfrac{2}{6}$$
$$\overline{\quad 5 \qquad \dfrac{1}{6} = 5\dfrac{1}{6}}$$

S. $4\dfrac{1}{4}$

   $-1\dfrac{1}{5}$
   —————

S. $5\dfrac{1}{3}$

   $-2\dfrac{1}{2}$
   —————

1. $3\dfrac{7}{8}$

   $-1\dfrac{1}{4}$
   —————

2. $9\dfrac{5}{6}$

   $-2\dfrac{1}{3}$
   —————

3. $5\dfrac{1}{4}$

   $-2\dfrac{2}{3}$
   —————

4. $7\dfrac{1}{8}$

   $-2\dfrac{1}{2}$
   —————

5. $2\dfrac{1}{7}$

   $-1\dfrac{3}{14}$
   —————

6. $9\dfrac{1}{4}$

   $-3\dfrac{7}{16}$
   —————

7. $6\dfrac{2}{3}$

   $-3\dfrac{4}{9}$
   —————

8. $6\dfrac{1}{2}$

   $-2\dfrac{2}{3}$
   —————

9. $7\dfrac{1}{4}$

   $-2\dfrac{3}{5}$
   —————

10. $7\dfrac{1}{8}$

    $-4\dfrac{3}{4}$
    —————

1	
2	
3	
4	
5	
6	
7	
8	
9	
10	
Score	

Problem Solving

There are 312 students enrolled in a school. If they have been placed into thirteen equal-sized classes, how many students are in each class?

Review Exercises

Notes

1. Reduce $\dfrac{24}{30}$ to its lowest terms

2. Convert $\dfrac{29}{4}$ to a mixed numeral

3. Find the least common denominator for the following fractions:
$$\dfrac{1}{3}, \ \dfrac{5}{6}, \text{ and } \dfrac{3}{4}$$

4.
$$\begin{array}{r} \dfrac{1}{3} \\[4pt] \dfrac{1}{5} \\[4pt] +\dfrac{1}{2} \\ \hline \end{array}$$

Helpful Hints

Use what you have learned to solve the following problems.

*Be sure all fractions are reduced to lowest terms.

S. $3\dfrac{1}{7}$ $-1\dfrac{5}{7}$

S. $6\dfrac{1}{2}$ $+3\dfrac{3}{4}$

1. $\dfrac{8}{9}$ $-\dfrac{1}{2}$

2. $\dfrac{8}{9}$ $-\dfrac{1}{6}$

3. 7 $-2\dfrac{3}{5}$

4. $5\dfrac{1}{8}$ $+3\dfrac{1}{2}$

5. $7\dfrac{7}{8}$ $+3\dfrac{3}{8}$

6. $7\dfrac{1}{2}$ $-2\dfrac{3}{4}$

7. $3\dfrac{4}{5}$ $+4\dfrac{2}{3}$

8. $6\dfrac{1}{2}$ -3

9. $\dfrac{7}{16}$ $+\dfrac{1}{4}$

10. $4\dfrac{5}{6}$ $+3\dfrac{3}{4}$

1	
2	
3	
4	
5	
6	
7	
8	
9	
10	
Score	

Problem Solving

Susan earned $3\dfrac{3}{4}$ dollars on Monday and $7\dfrac{1}{2}$ dollars on Tuesday. How much more did she earn on Tuesday than on Monday?

Review Exercises

Notes

1. $6 \overline{)726}$

2. $\begin{array}{r} 725 \\ \times\ 36 \\ \hline \end{array}$

3. $73 + 13 + 76 + 59 =$

4. $8{,}033 - 1{,}765 =$

Helpful Hints	When multiplying common fractions, first multiply the numerators. Next, multiply the denominators. If the answer is an improper fraction, change it to a mixed numeral.	**Examples:** $\frac{3}{4} \times \frac{2}{7} = \frac{6}{28} = \frac{3}{14}$ $\frac{3}{2} \times \frac{7}{8} = \frac{21}{16} = 1\frac{5}{16}$	*Be sure to reduce fractions to lowest terms.

S. $\dfrac{3}{4} \times \dfrac{5}{7} =$ S. $\dfrac{4}{5} \times \dfrac{3}{5} =$ 1. $\dfrac{2}{9} \times \dfrac{1}{7} =$ 2. $\dfrac{2}{5} \times \dfrac{1}{2} =$

3. $\dfrac{7}{2} \times \dfrac{3}{5} =$ 4. $\dfrac{7}{9} \times \dfrac{2}{3} =$ 5. $\dfrac{8}{9} \times \dfrac{3}{4} =$ 6. $\dfrac{4}{3} \times \dfrac{4}{5} =$

7. $\dfrac{3}{2} \times \dfrac{4}{5} =$ 8. $\dfrac{2}{7} \times \dfrac{3}{5} =$ 9. $\dfrac{1}{2} \times \dfrac{4}{5} =$ 10. $\dfrac{5}{2} \times \dfrac{3}{7} =$

1	
2	
3	
4	
5	
6	
7	
8	
9	
10	
Score	

Problem Solving	A family roasted $2\frac{1}{4}$ pounds of beef for dinner and ate $1\frac{3}{5}$ pounds. How much beef was left?

Review Exercises

Notes

1. $\dfrac{2}{3} \times \dfrac{6}{7} =$ 2. $\dfrac{7}{3} \times \dfrac{4}{5} =$

3. $\begin{array}{r} \dfrac{7}{8} \\ - \dfrac{1}{8} \\ \hline \end{array}$ 4. $\begin{array}{r} \dfrac{2}{3} \\ + \dfrac{2}{3} \\ \hline \end{array}$

Helpful Hints	If the numerator of one fraction and the denominator of another have a common factor, they can be divided out before you multiply the fractions.	**Examples:** 4 is a common factor $\dfrac{3}{1\cancel{4}} \times \dfrac{\cancel{8}^{2}}{11} = \dfrac{6}{11}$	2 is a common factor $\dfrac{7}{4\cancel{8}} \times \dfrac{\cancel{6}^{3}}{5} = \dfrac{21}{20} = 1\dfrac{1}{20}$

S. $\dfrac{3}{5} \times \dfrac{5}{7} =$ S. $\dfrac{9}{10} \times \dfrac{5}{3} =$ 1. $\dfrac{2}{5} \times \dfrac{15}{16} =$ 2. $\dfrac{8}{15} \times \dfrac{3}{16} =$

3. $\dfrac{5}{6} \times \dfrac{7}{15} =$ 4. $\dfrac{7}{3} \times \dfrac{10}{7} =$ 5. $\dfrac{5}{8} \times \dfrac{12}{25} =$ 6. $\dfrac{8}{9} \times \dfrac{3}{4} =$

7. $\dfrac{3}{4} \times \dfrac{8}{15} =$ 8. $\dfrac{3}{4} \times \dfrac{3}{5} =$ 9. $\dfrac{4}{3} \times \dfrac{6}{7} =$ 10. $\dfrac{5}{6} \times \dfrac{4}{7} =$

1	
2	
3	
4	
5	
6	
7	
8	
9	
10	
Score	

Problem Solving	There are 15 rows of seats in a theater. If each row has 26 seats, how many seats are there in the theater?

Review Exercises

Notes

1. $\dfrac{3}{5} \times \dfrac{15}{21} =$ 2. $\dfrac{8}{9} \times \dfrac{7}{12} =$

3. $\begin{array}{r} \dfrac{2}{3} \\[6pt] + \dfrac{1}{5} \\ \hline \end{array}$ 4. $\begin{array}{r} \dfrac{3}{4} \\[6pt] - \dfrac{2}{3} \\ \hline \end{array}$

| **Helpful Hints** | When multiplying whole numbers and fractions, write the whole number as a fraction and then multiply. | **Examples:** $\dfrac{2}{3} \times 15 =$ $\dfrac{2}{1\cancel{3}} \times \dfrac{\cancel{15}^{5}}{1} = \dfrac{10}{1} = 10$ | $\dfrac{3}{4} \times 9 =$ $\dfrac{3}{4} \times \dfrac{9}{1} = \dfrac{27}{4}$ | $4\overline{)27}\;\;6\tfrac{3}{4}$ $\dfrac{24}{3}$ |

S. $\dfrac{3}{4} \times 12 =$ S. $\dfrac{2}{3} \times 5 =$ 1. $\dfrac{3}{4} \times 8 =$ 2. $10 \times \dfrac{2}{5} =$

3. $\dfrac{4}{5} \times 25 =$ 4. $\dfrac{2}{7} \times 4 =$ 5. $\dfrac{1}{2} \times 27 =$ 6. $\dfrac{1}{10} \times 25 =$

7. $6 \times \dfrac{7}{12} =$ 8. $\dfrac{5}{6} \times 9 =$ 9. $\dfrac{3}{8} \times 40 =$ 10. $\dfrac{4}{5} \times 7 =$

1	
2	
3	
4	
5	
6	
7	
8	
9	
10	
Score	

Problem Solving	A class has 36 students. If $\dfrac{2}{3}$ of them are girls, how many girls are there in the class?

Review Exercises

Notes

1.　$63\overline{)796}$

2.　$\dfrac{3}{4} \times 16 =$

3.　$\dfrac{2}{3} \times 10 =$

4.　Change $3\dfrac{1}{2}$ to an improper fraction.

| **Helpful Hints** | To multiply mixed numerals, first change them to improper fractions, then multiply. Express answers in lowest terms. | **Example:** $1\dfrac{1}{2} \times 1\dfrac{5}{6} =$ $\dfrac{\cancel{3}}{2} \times \dfrac{11}{\cancel{6}_2} = \dfrac{11}{4} = 2\dfrac{3}{4}$ |

S.　$\dfrac{2}{3} \times 1\dfrac{1}{8} =$

S.　$1\dfrac{1}{4} \times 2\dfrac{2}{5} =$

1.　$\dfrac{3}{4} \times 2\dfrac{1}{2} =$

2.　$2\dfrac{1}{3} \times 1\dfrac{1}{3} =$

3.　$5 \times 3\dfrac{1}{5} =$

4.　$2\dfrac{1}{7} \times 1\dfrac{2}{5} =$

5.　$2\dfrac{2}{3} \times 2\dfrac{1}{4} =$

6.　$2\dfrac{1}{4} \times 1\dfrac{1}{2} =$

7.　$2\dfrac{1}{2} \times 3\dfrac{1}{4} =$

8.　$6 \times 2\dfrac{1}{2} =$

9.　$2\dfrac{1}{2} \times 4\dfrac{2}{3} =$

10.　$2\dfrac{1}{6} \times \dfrac{3}{5} =$

1	
2	
3	
4	
5	
6	
7	
8	
9	
10	
Score	

Problem Solving　If a man can run 4 miles in an hour, at this rate, how far can he run in $3\dfrac{1}{2}$ hours?

Review Exercises

Notes

1. $\dfrac{3}{5} \times \dfrac{4}{7} =$ 2. $\dfrac{3}{4} \times \dfrac{9}{25} =$

3. $\dfrac{3}{4} \times 24 =$ 4. $13 \times \dfrac{2}{3} =$

Helpful Hints | Use what you have learned to solve the following problems. | *Be sure to express answers in lowest terms.
*Sometimes common factors may be divided out before you multiply.

S. $\dfrac{3}{5} \times \dfrac{1}{2} =$ S. $3\dfrac{1}{2} \times 2\dfrac{1}{7} =$ 1. $\dfrac{4}{5} \times \dfrac{7}{8} =$ 2. $\dfrac{20}{21} \times \dfrac{7}{40} =$

3. $\dfrac{3}{5} \times 35 =$ 4. $\dfrac{4}{7} \times 9 =$ 5. $\dfrac{3}{4} \times 2\dfrac{1}{2} =$ 6. $3\dfrac{2}{3} \times \dfrac{1}{2} =$

7. $5 \times 3\dfrac{2}{5} =$ 8. $1\dfrac{2}{3} \times 1\dfrac{2}{5} =$ 9. $3\dfrac{1}{6} \times 4\dfrac{4}{5} =$ 10. $1\dfrac{7}{8} \times 4\dfrac{1}{3} =$

1	
2	
3	
4	
5	
6	
7	
8	
9	
10	
Score	

Problem Solving | If a factory can produce $4\dfrac{1}{2}$ tons of parts in a day, how many tons can it produce in 5 days?

Review Exercises

Notes

1. $$\frac{3}{5}$$
    $$+\frac{1}{3}$$
    _____

2. $$\frac{3}{4}$$
    $$-\frac{1}{2}$$
    _____

3. $2 \times 3\frac{1}{2} =$

4. $1\frac{1}{3} \times 1\frac{1}{3} =$

Helpful Hints

To find the reciprocal of a common fraction, invert the fraction. To find the reciprocal of a mixed numeral, change the mixed numeral to an improper fraction, then invert it. To find the reciprocal of a whole number, first make it a fraction then invert it.

Examples: The reciprocal of:

$\frac{3}{5}$ is $\frac{5}{3}$ or $1\frac{2}{3}$ $2\frac{1}{2}$ is $\frac{2}{5}$ 7 is $\frac{1}{7}$
                                           $(\frac{5}{2})$ $(\frac{7}{1})$

Find the reciprocal of each number:

1	
2	
3	
4	
5	
6	
7	
8	
9	
10	
Score	

S. $\frac{3}{4}$ S. $2\frac{1}{3}$ 1. 6   2. $\frac{7}{8}$

3. $3\frac{1}{4}$ 4. 13   5. $\frac{2}{5}$ 6. $\frac{1}{7}$

7. 9   8. $\frac{2}{9}$ 9. $4\frac{1}{2}$ 10. 15

Problem Solving

Five students earned 225 dollars. If they divided the money equally among themselves, how much did each student receive?

Review Exercises

Notes

1. Find the reciprocal of 9

2. Find the reciprocal of $\dfrac{2}{7}$

3. Find the reciprocal of $3\dfrac{2}{3}$

4. $2\dfrac{2}{3} \times 1\dfrac{1}{5} =$

Helpful Hints

To divide fractions, find the reciprocal of the second number, then multiply the fractions.

Examples:

$\dfrac{2}{3} \div \dfrac{1}{2} =$

$\dfrac{2}{3} \times \dfrac{2}{1} = \dfrac{4}{3} = \boxed{1\dfrac{1}{3}}$

$2\dfrac{1}{2} \div 1\dfrac{1}{2} = \dfrac{5}{2} \div \dfrac{3}{2} =$

$\dfrac{5}{2} \times \dfrac{2}{3} = \dfrac{5}{3} = \boxed{1\dfrac{2}{3}}$

S. $\dfrac{3}{7} \div \dfrac{3}{8} =$

S. $3\dfrac{1}{2} \div 2 =$

1. $\dfrac{3}{8} \div \dfrac{1}{6} =$

2. $\dfrac{1}{2} \div \dfrac{1}{3} =$

3. $5 \div \dfrac{2}{3} =$

4. $4\dfrac{1}{2} \div \dfrac{1}{2} =$

5. $1\dfrac{3}{4} \div \dfrac{3}{8} =$

6. $5\dfrac{1}{4} \div \dfrac{7}{12} =$

7. $1\dfrac{1}{2} \div 3 =$

8. $5\dfrac{1}{2} \div 2 =$

9. $7\dfrac{1}{2} \div 2\dfrac{1}{2} =$

10. $3\dfrac{2}{3} \div 2\dfrac{1}{2} =$

1	
2	
3	
4	
5	
6	
7	
8	
9	
10	
Score	

Problem Solving

$3\dfrac{1}{2}$ yards are to be divided into pieces that are $\dfrac{1}{2}$ yards long. How many pieces will there be?

1. $\begin{array}{r} \frac{3}{5} \\ + \ \frac{1}{5} \\ \hline \end{array}$

2. $\begin{array}{r} \frac{5}{6} \\ + \ \frac{3}{6} \\ \hline \end{array}$

3. $\begin{array}{r} \frac{2}{3} \\ + \ \frac{1}{5} \\ \hline \end{array}$

4. $\begin{array}{r} 3\frac{2}{3} \\ + \ 4\frac{5}{9} \\ \hline \end{array}$

5. $\begin{array}{r} 7\frac{3}{4} \\ + \ 2\frac{3}{8} \\ \hline \end{array}$

6. $\begin{array}{r} \frac{5}{8} \\ - \ \frac{1}{8} \\ \hline \end{array}$

7. $\begin{array}{r} 7\frac{2}{5} \\ - \ 2\frac{3}{5} \\ \hline \end{array}$

8. $\begin{array}{r} 7 \\ - \ 2\frac{3}{5} \\ \hline \end{array}$

9. $\begin{array}{r} 6\frac{3}{4} \\ - \ \frac{1}{2} \\ \hline \end{array}$

10. $\begin{array}{r} 9\frac{1}{3} \\ - \ 3\frac{2}{5} \\ \hline \end{array}$

11. $\frac{2}{3} \ \text{x} \ \frac{4}{7} \ =$

12. $\frac{12}{13} \ \text{x} \ \frac{3}{24} \ =$

13. $\frac{3}{4} \ \text{x} \ 36 \ =$

14. $\frac{7}{8} \ \text{x} \ 2\frac{1}{7} \ =$

15. $2\frac{1}{3} \ \text{x} \ 3\frac{1}{2} \ =$

16. $\frac{3}{4} \div \frac{1}{2} \ =$

17. $3\frac{1}{2} \div \frac{1}{2} \ =$

18. $3\frac{2}{3} \div 1\frac{1}{2} \ =$

19. $3\frac{3}{4} \div 1\frac{1}{8} \ =$

20. $6 \div 2\frac{1}{3} \ =$

1	
2	
3	
4	
5	
6	
7	
8	
9	
10	
11	
12	
13	
14	
15	
16	
17	
18	
19	
20	

Review Exercises

Notes

1. $136 + 927 + 813$

2. $\dfrac{3}{5} + \dfrac{2}{3}$

2. $\begin{array}{r} 1{,}394 \\ -\ 966 \\ \hline \end{array}$

4. $\dfrac{3}{4} - \dfrac{1}{6}$

Helpful Hints

$\underset{\text{ones}}{1} \ . \ \underset{\text{tenths}}{2}\ \underset{\text{hundredths}}{3}\ \underset{\text{thousandths}}{4}\ \underset{\text{ten-thousandths}}{5}\ \underset{\text{hundred-thousandths}}{6}\ \underset{\text{millionths}}{7}$

To read decimals first read the whole number. Next, read the decimal point as "and." Next, read the number after the decimal point and its place value.

Examples:
3.16 = three and sixteen hundredths
14.011 = fourteen and eleven thousandths
0.69 = sixty-nine hundredths

Write the following in words:

S. 2.6   S. 13.016   1. 0.73   2. 4.002

3. 132.6   4. 132.06   5. 72.6395   6. 0.077

7. 9.89   8. 6.003   9. 0.72   10. 1.666

1	
2	
3	
4	
5	
6	
7	
8	
9	
10	
Score	

Problem Solving

In a class of 35 students, $\frac{3}{5}$ of them are boys. How many boys are there in the class?

Review Exercises

Notes

1.   234
    x 36

2. $\dfrac{3}{4}$ x $\dfrac{8}{11}$ =

3. $2\dfrac{1}{2} \div \dfrac{1}{2}$ =

4. $3\dfrac{1}{3} \div 2$ =

Helpful Hints	When reading remember "and" means decimal point. The fraction part of a decimal ends in "th" or "ths." Be careful about placeholders.

Examples:
Four and eight tenths = 4.8
Two hundred one and six hundredths = 201.06
One hundred four ten-thousandths = .0104

Write each of the following as a decimal. Use the chart at the bottom to help.

S. Six and four hundredths

S. Three hundred six and fifteen hundredths

1. Nine and eight tenths

2. Forty-six and thirteen thousandths

3. Three hundred twenty-six ten-thousandths

4. Fifty and thirty-nine thousandths

5. Eight hundred-thousandths

6. Four millionths

7. Twelve and thirty-six ten thousandths

8. Sixteen and twenty-four thousandths

9. Twenty-three and five tenths

10. Two and seventeen thousandths

1	
2	
3	
4	
5	
6	
7	
8	
9	
10	
Score	

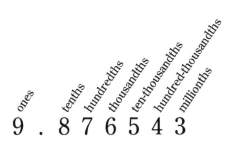

9 . 8 7 6 5 4 3

Problem Solving	If a normal temperature is $98\frac{3}{5}$ degrees, and a man has a temperature of 102 degrees, how much above normal is his temperature?

Review Exercises

Notes

1. $\dfrac{3}{4} \div \dfrac{1}{2} =$ 2. $1\dfrac{1}{2} \div 2 =$

3. $3 \div 1\dfrac{1}{2} =$ 4. $3\dfrac{1}{3} \div \dfrac{2}{5} =$

Helpful Hints	When changing mixed numerals to decimals, remember to put a decimal point after the whole number.	**Examples:** $3\dfrac{3}{10} = 3.3$ $\dfrac{16}{10,000} = .0016$ $42\dfrac{9}{10,000} = 42.0009$ $65\dfrac{12}{100,000} = 65.00012$

Write each of the following as a decimal. Use the chart at the bottom to help.

S. $7\dfrac{7}{10}$ S. $9\dfrac{7}{1,000}$ 1. $16\dfrac{32}{100}$ 2. $\dfrac{97}{10,000}$

3. $72\dfrac{9}{100}$ 4. $134\dfrac{92}{10,000}$ 5. $\dfrac{16}{1,000}$ 6. $44\dfrac{432}{100,000}$

7. $3\dfrac{96}{1,000}$ 8. $4\dfrac{901}{1,000}$ 9. $3\dfrac{901}{1,000,000}$ 10. $\dfrac{1,763}{100,000}$

ones . tenths hundredths thousandths ten-thousandths hundred-thousandths millionths

9 . 8 7 6 5 4 3

1	
2	
3	
4	
5	
6	
7	
8	
9	
10	
Score	

Problem Solving — A floor tile is $\dfrac{3}{4}$ inches thick. How many inches thick would a stack of forty-eight tiles be?

Review Exercises

Notes

1. $\dfrac{2}{3}$ 2. 7

    $+\dfrac{1}{5}$ $-1\dfrac{1}{3}$
    _____   _____

3. $\dfrac{2}{3} \times 1\dfrac{1}{2} =$ 4. $\dfrac{2}{3} \div 5\dfrac{1}{2} =$

Helpful Hints

Decimals can be easily changed to mixed numerals and fractions. Remember that the whole number is to the left of the decimal.

Examples: $2.6 = 2\dfrac{6}{10}$ $.210 = \dfrac{210}{1,000}$

$3.007 = 3\dfrac{7}{1,000}$ $1.0019 = 1\dfrac{19}{10,000}$

Change each of the following to a mixed numeral or fraction. Use the chart at bottom for help.

1	
2	
3	
4	
5	
6	
7	
8	
9	
10	
Score	

S. 1.43   S. 7.006   1. 173.016   2. .00016

3. 7.000014   4. 19.936   5. .09163   6. 77.8

7. 13.019   8. 72.0009   9. .00099   10. 63.000143

ones . tenths hundredths thousandths ten-thousandths hundred-thousandths millionths

9 . 8 7 6 5 4 3

Problem Solving

A theater has 9 rows of fourteen seats. Six of the seats are empty. How many of the seats are taken?

Review Exercises

Notes

1. Reduce $\frac{25}{30}$ to its simplest form.

2. Change $\frac{35}{8}$ to a mixed numeral.

3. Write 1.019 in words.

4. Change 72.008 to a mixed numeral.

Helpful Hints	Zeroes can be put to the right of a decimal without changing the value. This helps when comparing the value of decimals.	< means less than > means greater than	**Example:** Compare 4.3 and 4.28 4.3 =4.30 so 4.3 > 4.28

Place > or < to compare each pair of decimals.

S.   7.32 ☐ 7.6   S.   .99 ☐ .987   1.   6.096 ☐ 6.1

2.   3.41 ☐ 3.336   3.   7.11 ☐ 7.09   4.   1.5 ☐ 1.42

5.   3.62 ☐ .099   6.   .6 ☐ .79   7.   2.31 ☐ 2.4

8.   1.64 ☐ 1.596   9.   3.09 ☐ 3.4   10.   6.199 ☐ 6.2

1	
2	
3	
4	
5	
6	
7	
8	
9	
10	
Score	

Problem Solving

A group of hikers needed to travel 43 miles. They traveled 7 miles per day. After 5 days, how much farther did they still have to hike?

Review Exercises

Notes

1. $3\frac{1}{2} \div 2 =$ 2. $2\frac{1}{2} \div 2 =$

3.
$$\begin{array}{r} \frac{3}{8} \\ + \frac{5}{8} \\ \hline \end{array}$$

4.
$$\begin{array}{r} 7\frac{3}{5} \\ + 6\frac{4}{5} \\ \hline \end{array}$$

Helpful Hints	To add decimals, line up the decimal points and add as you would whole numbers. Write the decimal points in the answer. Zeroes may be placed to the right of the decimal.	**Example:** Add 3.16 + 2.4 + 12	$\begin{array}{r} 3.16 \\ 2.40 \\ + 12.00 \\ \hline 17.56 \end{array}$

S.
$$\begin{array}{r} 3.16 \\ 12.4 \\ + 3.26 \\ \hline \end{array}$$

S. $3.92 + 4.6 + .32 =$

1. $32.16 + 1.7 + 7.493 =$

2. $7.341 + 6.49 + .6 =$

3.
$$\begin{array}{r} 7.64 \\ 19.633 \\ + 2.4 \\ \hline \end{array}$$

4. $.37 + .6 + .73 =$

5. $9.64 + 7 + 1.92 + .7 =$ 6. $72.163 + 11.4 + 63.42 =$ 7. $.7 + .6 + .4 =$

8. $17.33 + 6.994 + .72 =$ 9.
$$\begin{array}{r} 7.642 \\ 17.63 \\ 2.143 \\ + 14.64 \\ \hline \end{array}$$

10. $19.2 + 7.63 + 4.26 =$

1	
2	
3	
4	
5	
6	
7	
8	
9	
10	
Score	

Problem Solving	In January it rained 3.6 inches, in February, 4.3 inches, and in March, 7.9 inches. What was the total amount of rainfall for the three months?

Review Exercises

Notes

1.   3.16
            3.4
       + 7.166

2.   3.6 + 4.16 + 8 =

3.   Find $\frac{1}{2}$ of $3\frac{1}{2}$

4.   Write $72\frac{9}{1,000}$ as a decimal.

Helpful Hints

To subtract decimals, line up the decimal points and subtract as you would whole numbers. Write the decimal points in the answer. Zeroes may be placed to the right of the decimal.

Examples:

$3.2 - 1.66 =$

$$\overset{2}{\cancel{3}}\overset{11}{.}\overset{1}{\cancel{2}}0$$
$$-\ 1.66$$
$$\overline{1.54}$$

$7 - 1.63 =$

$$\overset{6}{\cancel{7}}.\overset{9}{\cancel{0}}\overset{1}{0}$$
$$-\ 1.63$$
$$\overline{5.37}$$

1	
2	
3	
4	
5	
6	
7	
8	
9	
10	
Score	

S.   17.2
      −   3.36

S. 15.1 − 7.62 =

1.   7.32
       −   1.426

2.   3.962
      −   1.669

3. 2.72 − 1.56 =

4. 27.93 − 16.8 =

5. .72 − .667 =

6.   6.137
       − 2.1793

7. 3 − .627 =

8. 7.14 − 3.456 =

9. 75.6 − 66.972 =

10. 43.21 − 16.445 =

Problem Solving

Bill ran a race in 17.6 seconds and Jane ran it in 16.3 seconds. How much faster did Jane run the race than Bill?

Review Exercises

Notes

1. $\dfrac{1}{3}$
$+ \dfrac{1}{3}$
_____

2. $\dfrac{3}{8}$
$+ \dfrac{3}{8}$
_____

3. $\dfrac{7}{16}$
$+ \dfrac{13}{16}$
_____

4. $\dfrac{4}{5}$
$\dfrac{3}{5}$
$+ \dfrac{4}{5}$
_____

Helpful Hints	Use what you have learned to solve the following problems.	* Line up the decimal points. * Put decimal point in the answer. * Zeroes may be added to the right of the decimal point.

S. $\begin{array}{r} 3.61 \\ 14.4 \\ +\ \ .37 \\ \hline \end{array}$

S. $\begin{array}{r} 7.16 \\ -\ 3.473 \\ \hline \end{array}$

1. $\begin{array}{r} 7.16 \\ 8.92 \\ +\ 7.634 \\ \hline \end{array}$

2. $\begin{array}{r} 7.6 \\ -\ 1.43 \\ \hline \end{array}$

3. $4.63 + 5.7 + 6.24 =$

4. $17.2 - 8.96 =$

5. $15 - 12.92 =$

6. $6.93 + 5 + 7.63 =$

7. $.9 + .7 + .6 =$

8. $7.16 - 2.673 =$

9. $27.16 - 16.764 =$

10. $7.73 + 2.6 + .37 + 15 =$

1	
2	
3	
4	
5	
6	
7	
8	
9	
10	
Score	

Problem Solving	A worker earned $125.65. If he spent $76.93, how much of his earnings was left?

Review Exercises

Notes		
	1. 36 x 6	2. 46 x 32
	3. 209 x 23	4. 434 x 612

Helpful Hints

Multiply as you would with whole numbers. Find the number of decimal places and place the decimal point properly in the product.

Examples:

$$2.32 \leftarrow 2 \text{ places}$$
$$\underline{\text{x } \quad 6}$$
$$13.92 \leftarrow 2 \text{ places}$$

$$7.6 \leftarrow 1 \text{ place}$$
$$\underline{\text{x } 23}$$
$$228$$
$$\underline{1520}$$
$$174.8 \leftarrow 1 \text{ place}$$

								Score
S.	2.46 x 3	S.	2.3 x 16	1.	.643 x 3	2.	3.66 x 4	1
								2
								3
3.	.16 x 43	4.	.236 x 24	5.	1.4 x 16	6.	3.45 x 16	4
								5
								6
7.	7.63 x 43	8.	1.432 x 7	9.	.41 x 73	10.	.046 x 27	7
								8
								9
								10
								Score

Problem Solving

A hiker can travel 2.7 miles in an hour. At this pace, how far can the hiker travel in seven hours?

Review Exercises

Notes

1.   72.4
    x   3

2.   .27
        x   16

3. $\dfrac{3}{4}$ x $1\dfrac{1}{2}$ =

4. $2\dfrac{1}{8} \div 2 =$

Helpful Hints

Multiply as you would with whole numbers. Find the number of decimal places and place the decimal point properly in the product.

Examples:

$2.63 \leftarrow 2$ places
$\underline{x \ .3} \leftarrow 1$ place
$.789 \leftarrow 3$ places

$.724 \leftarrow 3$ places
$\underline{x \ .23} \leftarrow 2$ places
2172
$\underline{14480}$
$.16652 \leftarrow 5$ places

1	
2	
3	
4	
5	
6	
7	
8	
9	
10	
Score	

S.   3.6
    x .7

S.   3.24
    x 2.4

1.   3.6
    x 3.2

2.   2.09
    x .22

3.   .642
    x .33

4.   .23
    x 3.8

5.   2.03
    x .07

6.   .422
    x 23.2

7.   .003
    x 0.8

8.   5.6
    x .34

9.   63.5
    x 2.35

10.   12.3
    x .006

Problem Solving

A farmer can harvest 2.5 tons of potatoes in a day. How many tons can be harvested in 4.5 days?

Review Exercises

Notes

1. $\quad 6$
    $\quad - 2\frac{1}{3}$
    $\quad \overline{}$

2. $\quad 3\frac{1}{4}$
    $\quad - 1\frac{3}{4}$
    $\quad \overline{}$

3. $\quad \frac{2}{3}$
    $\quad + \frac{2}{3}$
    $\quad \overline{}$

4. $\quad 3\frac{1}{2}$
    $\quad + 4\frac{1}{2}$
    $\quad \overline{}$

Helpful Hints	To multiply by 10, move the decimal point one place to the right; by 100, two places to the right; by 1,000, three places to the right.	**Examples:** $\quad 10 \quad \times \quad 3.36 = 33.6$ $\quad 100 \quad \times \quad 3.36 = 336$ $\quad 1,000 \quad \times \quad 3.36 = 3360*$ *Sometimes placeholders are necessary

S. $10 \times 3.2 =$

S. $1,000 \times 7.39 =$

1. $100 \times .936 =$

2. $1,000 \times 72.6 =$

3. $100 \times 1.6 =$

4. $7.362 \times 100 =$

5. $7.28 \times 1,000 =$

6. $100 \times .7 =$

7. $100 \times .376 =$

8. $\quad 1,000$
   $\quad \times \quad .39$
   $\quad \overline{}$

9. $100 \times .733 =$

10. $10 \times 7.63 =$

1	
2	
3	
4	
5	
6	
7	
8	
9	
10	
Score	

Problem Solving

If tickets to a concert cost $9.50, how much would 1,000 tickets cost?

Review Exercises

Notes

1. $2\frac{1}{2} \div \frac{1}{3} =$

2. $3 \times 2\frac{1}{3} =$

3. $\frac{16}{17} \times \frac{7}{8} =$

4. $5 \div \frac{1}{2} =$

Helpful Hints

Use what you have learned to solve the following problems.   *Be careful when placing the decimal point in the product.

S. $\quad .342$ $\underline{\times\ 7}$	1
S. $\quad 42.3$ $\underline{\times\ .36}$	
1. $\quad .23$ $\underline{\times\ 14}$	2
	3
2. $\quad .29$ $\underline{\times\ 1.6}$	
3. $\quad 1.34$ $\underline{\times\ .362}$	4
4. $\ 10 \times 2.6 =$	
	5
5. $\quad 2.63$ $\underline{\times\ 1.2}$	6
6. $\ 100 \times 26.3 =$	
7. $\quad .003$ $\underline{\times\ 3.6}$	7
	8
8. $\quad .65$ $\underline{\times\ 5.5}$	
9. $\quad 1.67$ $\underline{\times\ 33}$	9
10. $\quad 0.67$ $\underline{\times\ .063}$	10
	Score

Problem Solving

Sweaters are on sale for $13.79. How much would three of the sweaters cost on sale?

63

Review Exercises

Notes

1. $7 \overline{)133}$ 2. $7 \overline{)1414}$

3. $6 \overline{)6006}$ 4. $22 \overline{)2442}$

Helpful Hints

Divide as you would with whole numbers. Place the decimal point directly up.

Examples:

$$\begin{array}{r} 2.8 \\ 3 \overline{)8.4} \\ -6\downarrow \\ \hline 24 \\ -24 \\ \hline 0 \end{array} \qquad \begin{array}{r} .084 \\ 3 \overline{).252} \\ -24\downarrow \\ \hline 12 \\ -12 \\ \hline 0 \end{array}$$

S. $3 \overline{)1.32}$ S. $8 \overline{)14.4}$ 1. $3 \overline{)59.1}$ 2. $7 \overline{)22.47}$

3. $34 \overline{)19.38}$ 4. $70.3 \div 19 =$ 5. $4 \overline{)24.32}$ 6. $6 \overline{)245.4}$

7. $26 \overline{)8.424}$ 8. $16 \overline{)2.56}$ 9. $12.72 \div 6 =$ 10. $21 \overline{)42.84}$

1	
2	
3	
4	
5	
6	
7	
8	
9	
10	
Score	

Problem Solving

A man has a board $10\frac{1}{2}$ feet long. If he decides to cut it into pieces $\frac{1}{2}$ foot long, how many pieces will there be?

Review Exercises

Notes

1. $3\overline{)6.54}$

2. $15\overline{)4.5}$

3. $3.63 + 12 + 3.2 =$

4. $\begin{array}{r} 7.2 \\ -\ 2.367 \\ \hline \end{array}$

Helpful Hints	Sometimes placeholders are necessary when dividing decimals.	**Examples:**	$\begin{array}{r} .05 \\ 3\overline{)\ .15} \\ -15 \\ \hline 0 \end{array}$	$\begin{array}{r} .003 \\ 15\overline{)\ .045} \\ -\ 45 \\ \hline 0 \end{array}$

S. $5\overline{)\ .0135}$ S. $13\overline{)\ .247}$ 1. $7\overline{)\ .0049}$ 2. $3\overline{)\ .036}$

3. $4\overline{)\ .224}$ 4. $13\overline{)\ .468}$ 5. $22\overline{)\ .946}$ 6. $9\overline{)\ .567}$

7. $52\overline{)\ 1.196}$ 8. $18\overline{)\ .396}$ 9. $9\overline{)\ .027}$ 10. $12\overline{)\ .816}$

1	
2	
3	
4	
5	
6	
7	
8	
9	
10	
Score	

Problem Solving	Five boys earned $27.55 doing yard work. If they decided to divide the money equally among themselves, how much would each receive?

Review Exercises

Notes

1.   11.4
         x   3

2.   .233
            x   .4

3. $4\frac{1}{2} \div 1\frac{1}{2} =$ 4. $\frac{2}{3} \div \frac{2}{9} =$

Helpful Hints

Sometimes zeroes need to be added to the dividend to complete the problem.

Examples:

$5\overline{)1.3}$

$15\overline{)2.7}$

$$\begin{array}{r} .26 \\ 5\overline{)1.30} \\ -1\,0\downarrow \\ \hline 30 \\ -30 \\ \hline 0 \end{array}$$

$$\begin{array}{r} .18 \\ 15\overline{)2.70} \\ -1\,5\downarrow \\ \hline 120 \\ -120 \\ \hline 0 \end{array}$$

S. $5\overline{)1.7}$ S. $25\overline{)1.5}$ 1. $2\overline{).13}$ 2. $5\overline{)3.1}$

3. $22\overline{)45.1}$ 4. $24\overline{)3.6}$ 5. $5\overline{)0.2}$ 6. $95\overline{)3.8}$

7. $20\overline{)2.4}$ 8. $4\overline{)6.3}$ 9. $5\overline{)0.3}$ 10. $5\overline{)2.09}$

1	
2	
3	
4	
5	
6	
7	
8	
9	
10	
Score	

Problem Solving

A girl was born in 1979. How old will she be in 2007?

Decimals

Review Exercises

Notes

1. $2\overline{)\,.15}$ 2. $5\overline{)\,.13}$

3. $100 \times 9.3 =$ 4. $1,000 \times 9.3 =$

Helpful Hints

Move the decimal point in the divisor the number of places necessary to make it a whole number. Move the decimal point in the dividend the same number of places.

Examples:

$$.3\overline{)\,2.4}\quad \begin{array}{r}.8\\ \underline{-2\,4}\\ 0\end{array}$$

$$.03\overline{)\,28.50}\quad \begin{array}{r}950.\,*\\ \underline{-27\downarrow}\\ 15\\ \underline{-15}\\ 0\end{array}$$

*Sometimes placeholders are necessary.

S. $.7\overline{)\,2.73}$ S. $.15\overline{)\,.036}$ 1. $.3\overline{)\,2.4}$ 2. $.03\overline{)\,5.1}$

3. $.9\overline{)\,.378}$ 4. $.04\overline{)\,3.2}$ 5. $.06\overline{)\,.324}$ 6. $2.1\overline{)\,6.72}$

7. $.26\overline{)\,.962}$ 8. $.18\overline{)\,.576}$ 9. $.04\overline{)\,2.3}$ 10. $.12\overline{)\,1.104}$

1	
2	
3	
4	
5	
6	
7	
8	
9	
10	
Score	

Problem Solving

If a car traveled 110.5 miles in two hours, what was its average speed per hour?

Review Exercises

Notes

1. $3\overline{)2.4}$ 2. $.03\overline{)1.5}$

3. $.5\overline{).19}$ 4. $.15\overline{).6}$

Helpful Hints

To change fractions to decimals, divide the numerator by the denominator. Add as many zeroes as necessary.

Examples:

$$\frac{3}{4} \quad 4\overline{)\begin{array}{l}.75 \\ 3.00 \\ -2\,8\!\downarrow \\ \hline 20 \\ -20 \\ \hline 0 \end{array}}$$

$$\frac{3}{8} \quad 8\overline{)\begin{array}{l}.375 \\ 3.000 \\ -2\,4\!\downarrow\!\downarrow \\ \hline 60 \\ -56 \\ \hline 40 \\ -40 \\ \hline 0 \end{array}}$$

Change each of the following fractions to decimals.

S. $\dfrac{1}{2}$ S. $\dfrac{5}{8}$ 1. $\dfrac{3}{5}$ 2. $\dfrac{1}{4}$

3. $\dfrac{2}{5}$ 4. $\dfrac{7}{8}$ 5. $\dfrac{11}{20}$ 6. $\dfrac{13}{25}$

7. $\dfrac{5}{8}$ 8. $\dfrac{4}{20}$ 9. $\dfrac{1}{5}$ 10. $\dfrac{7}{10}$

1	
2	
3	
4	
5	
6	
7	
8	
9	
10	
Score	

Problem Solving

A worker earned 60 dollars and put $\frac{1}{4}$ of it in a savings account. How much did the worker put into the savings account?

Review Exercises

Notes

1. $3.36 + 5 + 2.6 =$ 2. $3.2 - 1.63 =$

3. $\begin{array}{r} 3.12 \\ \times \ \ .7 \\ \hline \end{array}$ 4. Change $\dfrac{7}{8}$ to a decmial

Helpful Hints

Use what you have learned to solve the following problems.

* Add as many zeroes as necessary.
* Placeholders may be necessary.
* Place decimal points properly.

S. $7\overline{).035}$ S. $.06\overline{)2.4}$ 1. $3\overline{)2.28}$ 2. $5\overline{).37}$

3. $1.6\overline{).04}$ 4. $.3\overline{)1.35}$ 5. $.5\overline{).12}$ 6. $.005\overline{)1.42}$

7. $.04\overline{)1.324}$ 8. $2.1\overline{)34.02}$ 9. Change $\dfrac{2}{5}$ to a decmial   10. Change $\dfrac{5}{8}$ to a decmial

1	
2	
3	
4	
5	
6	
7	
8	
9	
10	
Score	

Problem Solving

John weighed 120.5 pounds in January. By June he had lost 3.25 pounds. How much did he weigh in June?

1.   3.72
     4.6
  + 3.963

2.   .3 + 2.96 + 7.1 =

3.   15.4 + 4 + 9.7 =

4.   37.3
   − 16.7

5.   7.1
   − 2.37

6.   6 − 1.43 =

7.   3.12
   x   3

8.   3.4
   x 16

9.   .47
   x 1.6

10.   .436
   x 3.21

11. 100 x 2.36 =

12. 1,000 x 2.7 =

13. 2 $\overline{)2.68}$

14. 5 $\overline{)7.3}$

15. .5 $\overline{).325}$

16. .003 $\overline{)1.2}$

17. .15 $\overline{).0075}$

18. 8.7 $\overline{).1131}$

19. Change $\dfrac{7}{8}$ to a decmial

20. Change $\dfrac{11}{25}$ to a decmial

1	
2	
3	
4	
5	
6	
7	
8	
9	
10	
11	
12	
13	
14	
15	
16	
17	
18	
19	
20	

Percents

Review Exercises

Notes

1. $\dfrac{3}{4} \div \dfrac{1}{2} =$

2. $\dfrac{2}{3} \times 4\dfrac{1}{2} =$

3. $\begin{array}{r} \dfrac{1}{2} \\ + \dfrac{2}{3} \\ \hline \end{array}$

4. $\begin{array}{r} \dfrac{2}{3} \\ - \dfrac{1}{5} \\ \hline \end{array}$

Helpful Hints

Percent means "per hundred" or "hundredths." If a fraction is expressed as hundredths, it can easily be written as a percent.

Examples:

$\dfrac{7}{100} = 7\%$ $\dfrac{3}{10} = \dfrac{30}{100} = 30\%$ $\dfrac{19}{100} = 19\%$

Change each of the following to percents:

S. $\dfrac{17}{100} =$ S. $\dfrac{9}{10} =$ 1. $\dfrac{6}{100} =$ 2. $\dfrac{99}{100} =$

3. $\dfrac{3}{10} =$ 4. $\dfrac{64}{100} =$ 5. $\dfrac{67}{100} =$ 6. $\dfrac{1}{100} =$

7. $\dfrac{7}{10} =$ 8. $\dfrac{14}{100} =$ 9. $\dfrac{80}{100} =$ 10. $\dfrac{62}{100} =$

1	
2	
3	
4	
5	
6	
7	
8	
9	
10	
Score	

Problem Solving

A woman bought four chairs. If each chair weighed $22\dfrac{1}{2}$ pounds, what was the total weight of the chairs?

Review Exercises

Notes

1. $\dfrac{7}{100}$ = _____ %

2. $\dfrac{9}{10}$ = _____ %

3. Find $\dfrac{1}{2}$ of $2\dfrac{1}{2}$

4. Find the difference between 17.6 and 9.85.

Helpful Hints	"Hundredths" = percent	**Examples:** .27 = 27% .9 = .90 = 90%
	Decimals can easily be changed to percents	*Move the decimal point twice to the right and add a percent symbol.

Change each of the following to percents:

				1	
S. .37	S. .7	1. .93	2. .02	2	
				3	
				4	
3. .2	4. .09	5. .6	6. .66	5	
				6	
				7	
7. .89	8. .6	9. .33	10. .8	8	
				9	
				10	
				Score	

Problem Solving

There are 32 fluid ounces in a quart. How many fluid ounces are there in .4 quarts?

Review Exercises

Notes

1. Reduce $\frac{18}{24}$ to its lowest terms.

2. Change $\frac{18}{16}$ to a mixed numeral with the fraction reduced to its lowest terms.

3. $5\frac{1}{5}$
    $-1\frac{1}{2}$
    _____

4. $2\frac{1}{2}$
    $+3\frac{3}{5}$
    _____

Helpful Hints	Percents can be expressed as decimals and as fractions. The fraction form may sometimes be reduced to its lowest terms.	**Examples:**	$25\% = .25 = \frac{25}{100} = \frac{1}{4}$ $8\% = .08 = \frac{8}{100} = \frac{2}{25}$

Change each percent to a decimal and to a fraction reduced to its lowest terms.

S. 20% = . ___ = ___   S. 9% = . ___ = ___   1. 16% = . ___ = ___

2. 6% = . ___ = ___   3. 75% = . ___ = ___   4. 40% = . ___ = ___

5. 1% = . ___ = ___   6. 45% = . ___ = ___   7. 12% = . ___ = ___

8. 5% = . ___ = ___   9. 50% = . ___ = ___   10. 13% = . ___ = ___

1	
2	
3	
4	
5	
6	
7	
8	
9	
10	
Score	

Problem Solving	75% of the students at Grover School take the bus. What fraction of the students take the bus? Reduce your fraction to its lowest terms.

Review Exercises

Notes

1. $.3 \overline{).54}$

2. Change $\dfrac{4}{5}$ to a decimal.

3.
$$\begin{array}{r} 3.12 \\ \times\ .6 \\ \hline \end{array}$$

4. $12 - 2.38 =$

Helpful Hints	To find the percent of a number, you may use either fractions or decimals. Use what is the most convenient.	**Example:** Find 25% of 60 .25 x 60	$\begin{array}{r} 60 \\ \times\ .25 \\ \hline 300 \\ 120 \\ \hline 15.00 \end{array}$ OR	$\dfrac{25}{100} = \dfrac{1}{4}$ $\dfrac{1}{4_1} \times \dfrac{60^{15}}{1} = \dfrac{15}{1} = 15$

S. Find 70% of 25.	S. Find 50% of 300.	1. Find 6% of 72.	**1**	
			2	
			3	
2. Find 60% of 85.	3. Find 25% of 60.	4. Find 45% of 250.	**4**	
			5	
5. Find 10% of 320.	6. Find 40% of 200.	7. Find 4% of 250.	**6**	
			7	
8. Find 90% of 240.	9. Find 75% of 150.	10. Find 2% of 660.	**8**	
			9	
			10	
			Score	

Problem Solving	A gasoline tank holds $3\frac{3}{4}$ gallons. If $\frac{1}{3}$ of the tank has been used, then how many gallons have been used?

Review Exercises

Notes

1. Find 13% of 85.

2. Change $\frac{4}{5}$ to a decimal.

3. Change 3% to a decimal.

4. Find 20% of 60.

When finding the percent of a number in a word problem, you can change the percent to a fraction or a decimal. Always express your answer in a short phrase or sentence.

Example:

A team played 60 games and won 75% of them. How many games did they win?

Find 75% of 60
.75 x 60

$$\begin{array}{r}60\\ \times\ .75\\ \hline 300\\ 420\\ \hline 45.00\end{array}$$

OR

$\frac{75}{100}=\frac{3}{4}$

$\frac{3}{4}\times\frac{60}{1}=\frac{45}{1}=45$

Answer: The team won 45 games.

Helpful Hints

S. George took a test with 20 problems. If he got 15% of the problems correct, how many problems did he get correct?

S. If 6% of the 500 students enrolled in a school are absent, then how many students are absent?

1. A worker earned 80 dollars and put 70% of it into the bank. How many dollars did he put into the bank?

2. A car costs $9,000. If Mr. Smith has saved 20% of this amount, how much did he save?

3. Steve took a test with 30 problems. If he got 70% of the problems correct, how many problems did he get incorrect?

4. A family's monthly income is $3,000. If 25% of this amount is spent on food, how many dollars are spent on food?

5. There are 40 students in a class. If 60% of the class are boys, then how many girls are in the class?

6. A house that costs $80,000 requires a 20% down payment. How many dollars are required for the down payment?

7. If a car costs $6,000 and loses 30% of its value in one year, how much will the car be worth in one year?

8. A coat is priced $50. If the sales tax is 7% of the price, how much is the sales tax? What is the total cost including sales tax?

9. 23% of the 600 students at Madison School take instrumental music. How many students are taking instrumental music?

10. A family spends 25% of its income for food and 30% for housing. If its monthly income is $3,000, how much is spent each month on food and housing?

1	
2	
3	
4	
5	
6	
7	
8	
9	
10	
Score	

Problem Solving

A train traveled 83.5 miles per hour. At this rate, how far would it travel in 2.5 hours?

Review Exercises

Notes

1. $20 \overline{)1764}$ 2. $25 \times 36 =$

3. $9 + 19 + 216 + 3{,}674 =$ 4. $7{,}010 - 6{,}914 =$

To change a fraction to a percent, first change the fraction to a decimal, then change the decimal to a percent. Move the decimal twice to the right and add a percent symbol.

Examples: $\dfrac{3}{4}$ $\begin{array}{r} .75 \\ 4\overline{)3.00} \\ -28 \\ \hline 20 \\ -20 \\ \hline 0 \end{array}$ $= 75\%$ $\dfrac{16}{20} = \dfrac{4}{5}$ $\begin{array}{r} .80 \\ 5\overline{)4.00} \\ -40 \\ \hline 0 \end{array}$ $= 80\%$

* Sometimes the fraction can be reduced further.

Helpful Hints

Change each of the following to percents:

S. $\dfrac{1}{5} =$ S. $\dfrac{12}{15} =$ 1. $\dfrac{3}{5} =$ 2. $\dfrac{1}{2} =$

3. $\dfrac{1}{10} =$ 4. $\dfrac{9}{12} =$ 5. $\dfrac{15}{20} =$ 6. $\dfrac{15}{25} =$

7. $\dfrac{1}{4} =$ 8. $\dfrac{24}{30} =$ 9. $\dfrac{18}{24} =$ 10. $\dfrac{4}{20} =$

1	
2	
3	
4	
5	
6	
7	
8	
9	
10	
Score	

Problem Solving

320 people applied for jobs at a new department store. If 20% of the people were given a job, how many people got jobs?

Review Exercises

Notes

1.   4.19
            x 3

2.   12.6
          − 3.743

3.   36.16
             .724
          + 7.93

4.   .05 ⌐ .235

When finding the percent, first write a fraction, change the fraction to a decimal, then change the decimal to a percent.

Examples:

4 is what percent of 16?

$\frac{4}{16} = \frac{1}{4}$

```
          .25 = 25%
    4 ⌐ 1.00
        - 8↓
          20
        - 20
           0
```

5 is what percent of 25?

$\frac{5}{25} = \frac{1}{5}$

```
          .20 = 20%
    5 ⌐ 1.00
        - 1 0↓
           00
```

Helpful Hints

S. 3 is what percent of 12?

S. 15 is what percent of 20?

1. 7 is what percent of 28?

2. 20 is what percent of 25?

3. 40 = what percent of 80?

4. 18 is what percent of 20?

5. 12 is what percent of 20?

6. 9 is what percent of 12?

7. 15 = what percent of 20?

8. 24 is what percent of 32?

9. 400 is what percent of 500?

10. 19 is what percent of 20?

1	
2	
3	
4	
5	
6	
7	
8	
9	
10	
Score	

Problem Solving

A man had 215 dollars in the bank. One day he withdrew 76 dollars and the next day he made a deposit of 96 dollars. How much does he now have in the bank?

Review Exercises

Notes	
	1. Find 12% of 220.

2. Change $\dfrac{3}{5}$ to a percent.

3. Find 60% of 45.

4. $.05 \overline{\smash{\big)}\,1.7}$

When finding the percent first write a fraction, change the fraction to a decimal, then change the decimal to a percent.

Helpful Hints

Example:

A team played 20 games and won 15 of them. What percent of the games did they win?

15 is what % of 20?

$\dfrac{15}{20} = \dfrac{3}{4}$

$.75 = 75\%$

$4\overline{\smash{\big)}\,3.00}$
$\underline{-\,28}$
20
$\underline{-\,20}$
0

They won 75% of the games.

S. A test had 20 questions. If Sam got 15 questions correct, what percent did he get correct?

S. In a class of 30 students, 12 are girls. What percent of the class is girls?

1. On a spelling test with 25 words, Susan got 20 correct. What percent of the words did she get correct?

2. A worker earned 200 dollars. If she put 150 dollars into a savings account, what percent of her earnings did she put into a savings account?

3. A team played 16 games and won 12 of them. What percent did they lose?

4. A quarterback threw 35 passes and 28 were caught. What percent of the passes were caught?

5. $\dfrac{19}{20}$ of a class was present at school. What percent of the class was present?

6. A class has an enrollment of 30 students. If 24 are present, what percent are absent?

7. A team won 12 games and lost 13 games. What percent of the games played did it win?

8. A school has 300 students. If 60 of them are sixth graders, what percent are sixth graders?

9. On a math test with 50 questions Jill got 49 of them correct. What percent did she get correct?

10. A pitcher threw 12 pitches. If 9 of them were strikes, what percent were strikes?

1	
2	
3	
4	
5	
6	
7	
8	
9	
10	
Score	

Problem Solving

There were 30 questions on a test. If a student got 80% of them correct, how many questions did he get correct?

Review Exercises

Notes

1. Change $\frac{7}{100}$ to a percent.   2. Change $\frac{9}{10}$ to a percent.

3. Change .3 to a percent.   4. Change 24% to a decimal and a fraction expressed in its lowest terms.

Helpful Hints

Use what you have learned to solve the following problems.

Examples:

Find 15% of 35
.15 x 35

$$\begin{array}{r} 35 \\ \times\ .15 \\ \hline 175 \\ 35\ \ \\ \hline 5.25 \end{array}$$

18 is what % of 24?

$\frac{18}{24} = \frac{3}{4}$

$$.75 = 75\%$$
$$4\overline{)3.00}$$
$$\begin{array}{r} -28 \\ \hline 20 \\ -20 \\ \hline 0 \end{array}$$

S. Find 20% of 45.	S. 3 is what percent of 12?	1. Find 3% of 120.

#	
1	
2	
3	
4	
5	
6	
7	
8	
9	
10	
Score	

S. Find 20% of 45.   S. 3 is what percent of 12?   1. Find 3% of 120.

2. Find 80% of 72.   3. 12 is what percent of 16?   4. 20 = what percent of 25?

5. Find 25% of 310.   6. Find 12% of 50.   7. 12 is what percent of 48?

8. 4 = what percent of 40?   9. Find 90% of 500.   10. Find 22% of 236.

Problem Solving

A worker has completed $\frac{4}{5}$ of his project. What percent of his project has been completed?

Review Exercises

Notes

1. $2\frac{1}{2}$ 2. $3\frac{1}{4}$

       $-\,2\frac{1}{3}$ $+\,2\frac{1}{2}$

3. $\dfrac{7}{10} \div \dfrac{3}{14} =$ 4. $\dfrac{4}{5} \div \dfrac{1}{3} =$

Use what you have learned to solve the following problems.   **Examples:**

A farmer has 210 cows. If he sells 40% of them, how many does he sell?

In a class of 24 students, 18 are girls. What percent are girls?

$$.75 = 75\%$$

40% of 210   210
.4 x 210   x .4   He sold 84 cows.
             84.0

18 is what % of 24? $\dfrac{18}{24} = \dfrac{3}{4}$

$$4\overline{)3.00}$$
$$\underline{-\,28}$$
$$20$$
$$\underline{-\,20}$$
$$0$$

75% are girls.

Helpful Hints

S. A test has 40 problems. A student got 80% of them correct. How many problems did he get correct?

S. Sue has finished 6 problems on a test. If there are 24 problems on the test, what percent has she finished?

1. A ranch has 500 acres of land. If 60% of the land is used for grazing, then how many acres are used for grazing?

2. A player took 15 shots. If he made 9 of them, what percent did he make?

3. A man earned 24 dollars and spent 60% of it. How much did he spend?

4. A test has 45 questions. If Jane got 36 correct, what percent did she get correct?

5. 27 is what percent of 36?

6. Find 24% of 60.

7. There are 400 students in a school. If 60% eat cafeteria food, how many students eat cafeteria food?

8. A baseball team played 20 games and won 18. What percent did they lose?

9. A car costs $6,000. If a down payment of 20% is required, how much is the down payment?

10. 60 players tried out for a team. If only 12 made the team, what percent made the team? What percent didn't make the team?

1	
2	
3	
4	
5	
6	
7	
8	
9	
10	
Score	

Problem Solving

A student's test scores were 84, 96, 80, and 76. What was the student's average test score?

Change numbers 1 through 5 to a percent.

1. $\dfrac{17}{100}$ =   2. $\dfrac{3}{100}$ =   3. $\dfrac{7}{10}$ =   4. .19 =   5. .6 =

Change numbers 6 through 8 to a decimal and a fraction expressed in lowest terms.

6. 9% = . = ____   7. 14% = . = ____   8. 80% = . = ____

Solve the following problems. Label the word problem answers.

9. Find 4% of 320.

10. Find 60% of 230.

11. Find 12% of 600.

12. 3 is what percent of 5?

13. 12 is what percent of 15?

14. 12 is what percent of 48?

15. Change $\dfrac{1}{5}$ to a percent.

16. Change $\dfrac{1}{4}$ to a percent.

17. A man earned 200 dollars. If he put 40% of it into the bank, how many dollars did he put into the bank?

18. If a rancher has 450 cows and decides to sell 40% of them, how many cows will he sell?

19. A class has 40 students enrolled. If 18 are boys, what percent are boys?

20. A pitcher threw 80 pitches and 60 were strikes. What percent were strikes?

1	
2	
3	
4	
5	
6	
7	
8	
9	
10	
11	
12	
13	
14	
15	
16	
17	
18	
19	
20	

Review Exercises

Notes

1. 3.6 + .72 + 3.9 =   2. 16.1 — 2.96 =

3.   1.64   4.   5 ⟌ 3.0
       x .03

Helpful Hints

Geometric term:	Point	Line	Plane	Line Segment	Ray
Example:	• P	A B		A B	A B
Symbol:	P	$\overleftrightarrow{AB}$	plane ABC	$\overline{AB}$	$\overrightarrow{AB}$

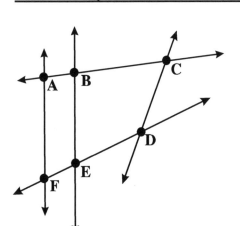

Use the figure to answer the following:

S. Name 4 points   S. Name 5 line segments

1. Name 5 lines   2. Name 5 rays

3. Name 3 points on   4. Give another name for
   $\overleftrightarrow{FD}$ $\overleftrightarrow{AB}$

5. Give another name for   6. Give another name for
   $\overleftrightarrow{ED}$ $\overrightarrow{AC}$

7. Name 2 line segments on $\overleftrightarrow{FD}$ 8. Name 2 rays on $\overleftrightarrow{FE}$

9. Name 2 rays on $\overleftrightarrow{AC}$ 10. What point is common to lines $\overleftrightarrow{FD}$ and $\overleftrightarrow{BE}$?

1	
2	
3	
4	
5	
6	
7	
8	
9	
10	
Score	

Problem Solving

If a factory can manufacture an engine in $2\frac{1}{2}$ hours, how long will it take to manufacture 10 engines?

Review Exercises

Notes

1. $\dfrac{3}{5}$

   $+\dfrac{4}{5}$
   _____

2. $\dfrac{3}{4}$

   $-\dfrac{1}{4}$
   _____

3. $2 \times \dfrac{3}{4} =$

4. $3 \div \dfrac{3}{4} =$

Helpful Hints

Geometric term: Parallel Lines   Intersecting Lines   Perpendicular Lines   Angle   Symbols
Example:

∠DAC
∠CAD
∠A

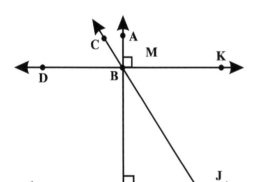

Use the figure to answer the following:

S. Name 2 parallel lines

S. Name 2 perpendicular lines

1. Name 3 pairs of intersecting lines

2. Name 5 angles

3. Name 3 angles that have B as their vertex

4. Name 3 angles that have H as their vertex

5. Name 3 lines   6. Name 5 line segments   7. Name 5 rays

8. Name 3 line segments on   9. Name 3 lines which include point B

10. Give another name for ∠JHI

1	
2	
3	
4	
5	
6	
7	
8	
9	
10	
Score	

Problem Solving

Mary has a piece of cloth $7\frac{1}{2}$ yards long. How many pieces of $1\frac{1}{2}$ yards long can she cut from this piece?

Review Exercises

Notes

1. 6 is what % of 8?   2. Find 15% of 225.

3. A man had 300 cows and
    decided to sell 15% of them.
    How many cows did he sell?

4. Sue took a test with 20
    problems. If she got 14 of the
    problems correct, then what
    percent of the problems did she
    get correct?

Helpful Hints

right angle — measures 90°   acute angle — measures less than 90°   obtuse angle — measures more than 90°   straight angle — measures 180°

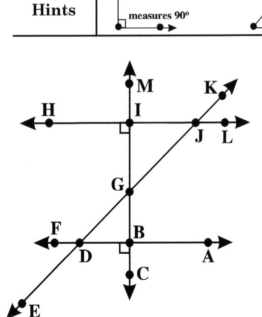

Use the figure to answer the following:

S. Name 4 right angles

S. Name 5 acute angles

1. Name 5 obtuse angles

2. Name 5 straight angles

3. What kind of angle is ∠IJG?

4. What kind of angle is ∠EDB?

5. What kind of angle is ∠GBD?

6. What kind of angle is ∠GJK?

7. Name an acute angle which has J as
    its vertex.

8. Name an obtuse angle which has D as
    its vertex.

9. Name a right angle which has B as
    its vertex.

10. Name a straight angle which has D as
    its vertex.

1	
2	
3	
4	
5	
6	
7	
8	
9	
10	
Score	

Problem Solving

A rope is 1.7 meters long. If a man wants to cut it into 5 pieces of
equal length, how long will each piece be?

Review Exercises

Notes

1. Change $\dfrac{9}{7}$ to a mixed numeral

2. Change $7\dfrac{2}{3}$ to an improper fraction

3. Express $\dfrac{24}{40}$ in its lowest terms

4. $2\dfrac{1}{2} \times 3\dfrac{1}{2} =$

Helpful Hints

To use a protractor, following these rules:
1. Place the center point of the protractor on the vertex.
2. Place the zero mark on one edge of the angle.
3. Read the number where the other side of the angle crosses the protractor.
4. If the angle is acute, use the smaller number. If the angle is obtuse, use the larger number.

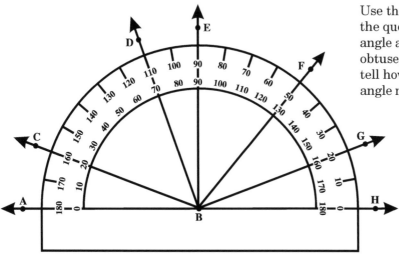

Use the figure to answer the questions. Classify the angle as right, acute, obtuse, or straight. Then tell how many degrees the angle measures.

1	
2	
3	
4	
5	
6	
7	
8	
9	
10	
Score	

S. ∠HBG   S. ∠DBH   1. ∠EBH   2. ∠CBH   3. ∠GBH   4. ∠DBA

5. ∠ABF   6. ∠FBH   7. ∠ABH   8. ∠ABG   9. ∠EBA   10. ∠FBA

Problem Solving

260 students took a social studies test and 80% received a passing grade. How many students received a passing grade?

Review Exercises

Notes

1. Change $\frac{15}{20}$ to a percent.

2. Change .9 to a percent.

3. Find 4% of 65.

4. 12 is what percent of 30?

Helpful Hints	When using a protractor, remember to follow these tips: 1. Place the center point of the protractor on the vertex of the angle. 2. Place the zero edge on the edge of the angle. 3. Read the number where the other side of the angle crosses the protractor. 4. If the angle is acute, use the smaller number. If the angle is obtuse, use the larger number.

Classify each angle as acute, right, obtuse or straight.

1	
2	
3	
4	
5	
6	
7	
8	
9	
10	
Score	

S. ∠AMC   S. ∠EMC   1. ∠DMA   2. ∠FMJ

3. ∠FMA   4. ∠DMJ   5. ∠EMA   6. ∠EMC

7. ∠FMB   8. ∠HMC   9. ∠IMA   10. ∠IMF

Problem Solving	There are 645 seats in an auditorium. If 379 of the seats are occupied, how many seats are empty?

Review Exercises

Notes

1. Name and classify this angle.   2. Name and classify this angle.

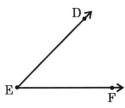

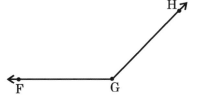

3. Name and classify this angle.   4. What kind of lines are these?

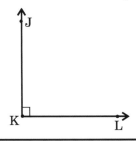

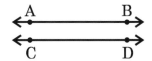

| **Helpful Hints** | Polygons are closed figures made up of line segments. | triangle 3 sides | rectangle 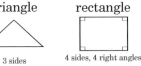 4 sides, 4 right angles | square 4 congruent sides, 4 right angles | parallelogram 4 sides, opposite sides parallel | trapezoid 4 sides, 1 pair of parallel sides |

Name each polygon. Some have more than one name.

S.   S.   1.   2.

   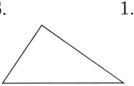

1	
2	
3	
4	
5	
6	
7	
8	
9	
10	
Score	

3.   4.   5.   6.

         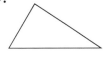

7.   8.   9.   10.

      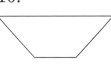

Problem Solving	A man earned $3.75 per hour. How much was his pay if he worked 8 hours?

Review Exercises

Notes

1. 30 ⟌ 7013

2. 48
    x 36

3. 732
    46
    + 377

4. 7611
    − 799

Helpful Hints

Triangles can be classified by sides and angles.

	Sides			Angles	

equilateral — 3 congruent sides
scalene — no congruent sides
isosceles — 2 congruent sides
acute — 3 acute angles
right — 1 right angle
obtuse — 1 obtuse angle

Classify each triangle by its sides and angles.

S. sides: _____ angles: _____

S. sides: _____ angles: _____

1. sides: _____ angles: _____

2. sides: _____ angles: _____

3. sides: _____ angles: _____

4. sides: _____ angles: _____

5. sides: _____ angles: _____
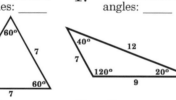

6. sides: _____ angles: _____

7. sides: _____ angles: _____
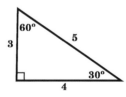

8. sides: _____ angles: _____
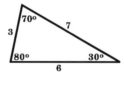

9. sides: _____ angles: _____
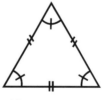

10. sides: _____ angles: _____
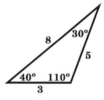

1	
2	
3	
4	
5	
6	
7	
8	
9	
10	
Score	

Problem Solving

Buses hold 60 people. How many buses are needed for 143 people?

Review Exercises

Notes

1. Classify by sides.

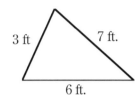
3 ft.   7 ft.
6 ft.

2. Classify by angles.

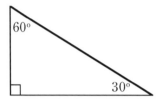
60°
30°

3. Classify by sides and angle.

sides: _____
angles: _____

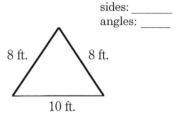
8 ft.   8 ft.
10 ft.

4. Find 60% of 75.

Helpful Hints | The distance around a polygon is its perimeter. | **Examples:**

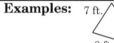 7 ft. 7 ft.
8 ft.
7
7
+ 8
perimeter = 22 ft.

 6 ft.
6
x 4
perimeter = 24 ft.

 4 ft.
6 ft.
2 x (6 + 4) =
2 x (10) =
perimeter = 20 ft.

Find the perimeter of each of the following.

S.
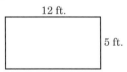
12 ft.
5 ft.

S.
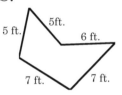
5 ft.   5ft.
6 ft.
7 ft.   7 ft.

1.

10 ft.
8 ft.   11 ft.
18 ft.

2.
12 ft.

3.
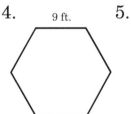
12 ft.   14 ft.
7 ft.

4.
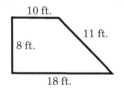
9 ft.

5.
13 ft.
22 ft.

6.
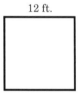
10 ft.
8 ft.   8 ft.
15 ft.

7.
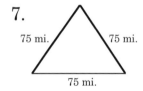
75 mi.   75 mi.
75 mi.

8.
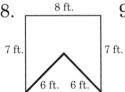
8 ft.
7 ft.   7 ft.
6 ft.   6 ft.

9.
21 ft.
22 ft.

10.
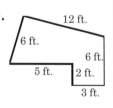
12 ft.
6 ft.
6 ft.
5 ft.   2 ft.
3 ft.

1	
2	
3	
4	
5	
6	
7	
8	
9	
10	
Score	

Problem Solving | A yard is in the shape of a rectangle which is 40 ft. wide and 55 ft. long. How many feet of fence will it take to go all the way around the yard?

89

Review Exercises

Notes

1. Find the perimeter.

7 ft. | 16 ft.

2. Find the perimeter.

14 ft. □

3. 6 is what % of 30?

4. $\dfrac{3}{4}$

   $-\dfrac{1}{3}$

   _____

Helpful Hints

These are the parts of a circle.

* The length of the diameter is twice that of the radius.

Use the figure to answer the following:

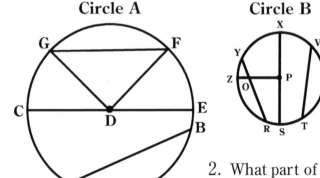

Circle A

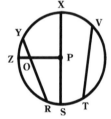

Circle B

S. What part of the circle is $\overline{CE}$?

S. Name 2 chords in circle B.

1. What part of circle A is $\overline{DF}$?

2. What part of circle B is $\overline{VT}$?

3. Name 3 radii in Circle A.

4. Name 2 chords in circle A.

5. If the length of $\overline{CE}$ is 16 ft., what is the length of $\overline{CD}$?

6. Name the center of Circle B.

7. Name 2 chords in circle B.

8. If $\overline{PS}$ in Circle B is 24 ft., what is the length of $\overline{XS}$?

9. Name 2 radii in Circle B.

10. Name a diameter in Circle B.

1	
2	
3	
4	
5	
6	
7	
8	
9	
10	
Score	

Problem Solving

A city is in the shape of a square. If its perimeter is 64 miles, what is the length of each side of the city?

Review Exercises

Notes

1. 325 + 16 + 9 =

2. 3 x 4.27 =

3.   .3 | 1.23

4. 7.6 + 14 + .3 + 2.14 =

The distance around a circle is called its circumference. The Greek letter Π = pi = 3.14 or $\frac{22}{7}$. To find the circumference, multiply Π x diameter. Circumference = Π x d

Examples:

C = Π x d
= 3.14 x 6

3.14
x 6
18.84 ft.

C = Π x d

= $\frac{22}{7_1}$ x $\frac{\cancel{14}^2}{1}$ = 44 ft.

(Hint: If the diameter is divisible by 7, use Π = $\frac{22}{7}$)

Helpful Hints

Find the circumference of each of the following. If there is no figure, draw a sketch.

S. 4 ft.

S. 10 ft.

1. 6 ft.

1	
2	
3	
4	
5	
6	
7	
Score	

2. 4 ft.

3. A circle with diameter 9 ft.

4. A circle with radius 14 ft.

5. 12 ft.

6. 5 ft.

7. A circle with radius 2 ft.

Problem Solving

A garden is in the shape of a circle. If the radius is 12 feet, how far is it all the way around the garden?

Review Exercises

Notes

1.   3.2
       x 6.1

2. $\dfrac{3}{4}$ x $1\dfrac{1}{3}$ =

3. $2\dfrac{1}{2}$ x $1\dfrac{1}{5}$ =

4. 7 x 32.5 =

The number of square units needed to cover a region is called its area.

Examples:

area square = side x side

$$A = s \times s$$
$$A = 7 \times 7$$

s = 7 ft.

   7
x 7
49 sq. ft.

area rectangle = length x width

w = 7 ft.

l = 12 ft.

$$A = l \times w$$
$$A = 12 \times 7$$

  12
x 7
84 sq. ft.

Hint: 1. Start with formulas   2. Substitute values   3. Solve the problem

Helpful Hints

Find the following areas.

S. [13 ft.]

S. [15 ft. / 11 ft.]

1. [14 ft. / 6 ft.]

2. [20 ft.]

3. A rectangle with length 12 ft. and width 11 ft.

4. [2.5 ft. / 4.3 ft.]

5. [$4\dfrac{1}{2}$ ft.] $1\dfrac{1}{3}$ ft.

6. [25 ft.]

7. A square with sides $2\dfrac{1}{2}$ ft.

1	
2	
3	
4	
5	
6	
7	
Score	

Problem Solving

A floor is the shape of a rectangle. The length is 14 feet and the width is 13 feet. What is the area of the floor?

Review Exercises

Notes

1. Find the area

16 ft.

14 ft.

2. Find the area

16 ft.

3. Find the circumference

8 ft.

4. Find the circumference

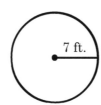

7 ft.

Area of a triangle = $\dfrac{\text{base x height}}{2} = \dfrac{\text{b x h}}{2}$

Area parallelogram = base x height = b x h

Examples:

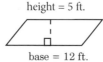

$A = \dfrac{\text{b x h}}{2}$

height = 8 ft.

base = 7 ft.

$A = \dfrac{7 \text{ x } 8}{2} = \dfrac{56}{2}$

28 sq. ft.

2 ⟌ 56

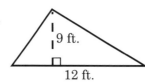

height = 5 ft.

base = 12 ft.

$A = \text{b x h}$
$A = 12 \text{ x } 5$

12
x 5
60 sq. ft.

Helpful Hints

Find the following areas.

S.

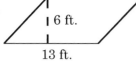

6 ft.

13 ft.

S.

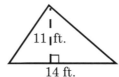

11 ft.

14 ft.

1.

9 ft.

12 ft.

1	
2	
3	
4	
5	
6	
7	
Score	

2.

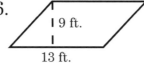

11 ft.

16 ft.

3. A triangle with base 5 ft. and height 7 ft.

4. A parallelogram with base 13 ft. and height 7 ft.

5.

14 ft.

12 ft.

6.

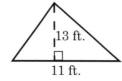

9 ft.

13 ft.

7.

13 ft.

11 ft.

Problem Solving

A circular dodge-ball court has a diameter of 30 feet. How many feet is it around the dodge-ball court?

Review Exercises

Notes

1. Find the area

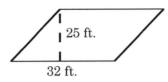

25 ft.

32 ft.

2. Find the area

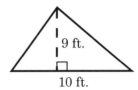

9 ft.

10 ft.

3. Find the area

7 ft.

14 ft.

4. Find the area

13 ft.

Area Circle = Π x radius x radius   A = Π x r x r

If the radius is divisible by 7, use Π = $\frac{22}{7}$

Helpful Hints

Examples:

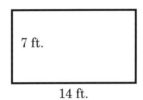

A = Π x r x r
= 3.14 x 3 x 3
= 3.14 x 9

3.14
x 9
28.26 sq. ft.

A = Π x r x r
= $\frac{22}{7_1}$ x $\frac{7^1}{1}$ x $\frac{7}{1}$
= 22 x 7

22
x 7
154 sq. ft.

Find the area of each circle.

S.
4 ft.

S.
12 ft.

1.
5 ft.

2.
14 ft.

3.
2 ft.

4.
8 ft.

5.
10 ft.

6. Circle with radius 6 ft.

7. Circle with diameter 14 ft.

1	
2	
3	
4	
5	
6	
7	
Score	

Problem Solving

A garden is in the shape of a circle. If the radius is 12 feet, how far is it all the way around the garden?

Review Exercises

Notes

1. Find the perimeter

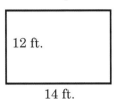

12 ft.

14 ft.

2. Find the area

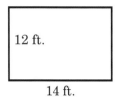

12 ft.

14 ft.

3. Find the circumference

6 ft.

4. Find the area

6 ft.

Helpful Hints

Remember these formulas.   For Areas: 1. Write formula 2. Substitute values 3. Solve problem

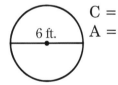

$$A = \frac{b \times h}{2}$$

P = Sum of 4 sides

A = b x h

C = Π x d   P = 4 x s   P = 2 (l + w)   P = Sum of all sides
A = Π x r x r   A = s x s   A = l x w

Find the perimeter or circumference. Second, find the area.

S.
12 ft.
7 ft.

P =   A =

S.

8 ft.   7 ft.   P =
        l 5 ft.   A =
        8 ft.

1.
12 ft.

P =
A =

2.
10 ft.
12 ft.

P =
A =

3.
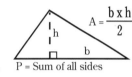
7 ft.   l 6 ft.
        l
        12 ft.

P =
A =

4.
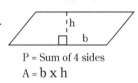
6 ft.

C =
A =

5.
14 ft.

C =
A =

6.
10 ft.
6 ft.
8 ft.

P =
A =

7. A square with sides 8 ft.

P =   A =

1	
2	
3	
4	
5	
6	
7	
Score	

Problem Solving

A man wants to buy a tent which is in the shape of a rectangle. If the length is 18 feet and the width is 12 feet, how many square feet of canvas will it take to make the tent?

Review Exercises

Notes

1.
$$\frac{3}{5}$$
$$-\frac{1}{2}$$
_____

2.
$$\frac{2}{3}$$
$$+\frac{1}{2}$$
_____

3. $3\frac{1}{2}$ x 3 =

4. $2\frac{1}{2} \div \frac{1}{2}$ =

Helpful Hints

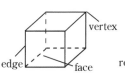 vertex 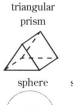 cube triangular prism triangular pyramid 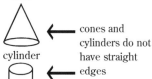 cone

edge face rectangular prism sphere square pyramid cylinder

cones and cylinders do not have straight edges

Identify the shape and the numbers of each part.

S.
name _____
faces _____
edges _____
vertices _____

S.
name _____
faces _____
edges _____
vertices _____

1.
name _____
faces _____
edges _____
vertices _____

2.
name _____
faces _____
edges _____
vertices _____

3.
name _____
faces _____
edges _____
vertices _____

4.
name _____
faces _____
edges _____
vertices _____

5.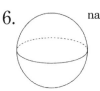
name _____
faces _____
edges _____
vertices _____

6.
name _____

7. How many more faces does a cube have than a triangular prism?

1	
2	
3	
4	
5	
6	
7	
Score	

Problem Solving

In a class of 40 people, 16 are girls. What percent of the class are girls? What percent are boys?

Use the figure to answer questions 1 - 8

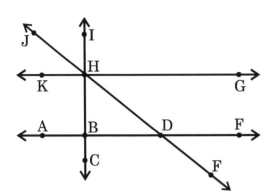

1. Name 2 parallel lines
2. Name 2 perpendicular lines
3. Name 4 line segments
4. Name 4 rays
5. Name 2 acute angles
6. Name 2 obtuse angles
7. Name 1 straight angle
8. Name 2 right angles

Triangle A

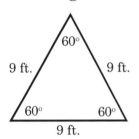

Triangle B

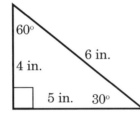

Use the figures to answer 9 and 10

9. Classify Triangle A by its sides and angles.

10. Classify Triangle B by its sides and angles.

11. Find the Perimeter

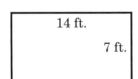

12. Find the Circumference

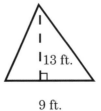

13. Find the Area

14. Find the Area

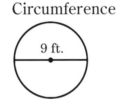

15. Find the Area

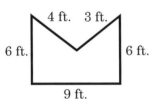

16. Find the Area

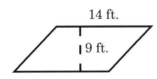

17. Find the Area

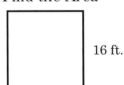

18. Identify and count the number of faces, edges, and vertices

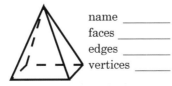

name _____
faces _____
edges _____
vertices _____

19. Find the Area

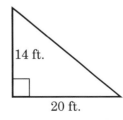

20. Find the perimeter of a square with sides of 96 feet.

	1
	2
	3
	4
	5
	6
	7
	8
	9
	10
	11
	12
	13
	14
	15
	16
	17
	18
	19
	20

Review Exercises

Notes

1. Find the area.

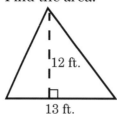

12 ft.

13 ft.

2. Find the circumference

13 ft.

3. 12 is what % of 16?

4. Find 3% of 425

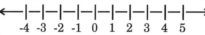

-4 -3 -2 -1 0 1 2 3 4 5

Integers to the left of zero are negative and less than zero. Integers to the right of zero are positive and greater than zero. When two integers are on a number line, the one farthest to the right is greater. **Hint:** Always find the sign of the answer first.

Helpful Hints

Examples: The sum of two negatives is a negative.

When adding a negative and a positive, the sign is the same as the integer farthest from zero. Then subtract.

-7 + -5 = -
(the sign is negative)

$$\begin{array}{r} 7 \\ + 5 \\ \hline 12 \end{array} = \boxed{-12}$$

-7 + 9 = +
(the sign is positive)

$$\begin{array}{r} 9 \\ - 7 \\ \hline 2 \end{array} = \boxed{+2}$$

1	
2	
3	
4	
5	
6	
7	
8	
9	
10	
Score	

S. -9 + 12 =

S. -15 + -6 =

1. -15 + 29 =

2. -12 + -6 =

3. 42 + -56 =

4. -15 + -16 =

5. -8 + 32 =

6. -39 + 76 =

7. -96 + -72 =

8. 73 + -86 =

9. -15 + -19 =

10. 72 + -81 =

Problem Solving

560 students attend Lincoln School. If 40% of them ride the bus to school, then how many students don't ride the bus to school?

Review Exercises

Notes

1. -16 + 18 =

2. 16 + -18 =

3. -16 + -18 =

4. Find the area.

3 ft.

Helpful Hints	When adding more than two integers, group the negatives and positives separately, then add.	**Examples:**

Examples:

$-6 + 4 + -5 =$ 11
$-11 + 4 = -$ - 4
(sign is negative)   7 = -7

$7 + -3 + -8 + 6 =$ 13
$-11 + 13 = +$ - 11
(sign is positive)   2 = +2

S. -3 + 5 + -6 =

S. -7 + 6 + -9 + 3 =

1. -3 + -4 + 5 =

2. 7 + -6 + -8 =

3. -15 + 19 + -12 =

4. -6 + 9 + 7 + 4 =

5. -16 + 32 + -18 =

6. -13 + 16 + -8 + 15 =

7. -9 + -7 + -6 =

8. -3 + 7 + -8 + -9 =

9. -32 + 16 + -17 + 8 =

10. -76 + 25 + -33 =

1	
2	
3	
4	
5	
6	
7	
8	
9	
10	
Score	

Problem Solving	The Giants played 36 games and won 27. What percent did they win? What percent did they lose?

Review Exercises

Notes

1. Find the perimeter of a rectangle with length 29 inches and width 13 inches.

2. -3 + 7 + -6 + 3 =

3. -12 + 27 + -6 =

4. -14 + -12 + 6 + 17 =

Helpful Hints	To subtract integers means to add the opposite.	**Examples:**			
		-3 - -8 = $\quad$ 8 -3 + 8 = + $\quad$ - 3 (sign is positive) $\overline{5 = (+5)}$	8 - 10 = $\quad$ 10 8 + -10 = - $\quad$ - 8 (sign is negative) $\overline{2 = (-2)}$	6 - -7 = $\quad$ 7 6 + 7 = + $\quad$ + 6 (sign is positive) $\overline{13 = (+13)}$	

			#	
S. -6 - 8 =	S. -6 - 9 =	1. 3 - -9 =	1	
			2	
			3	
2. 15 - 18 =	3. -16 - -25 =	4. -16 - 12 =	4	
			5	
5. 32 - -14 =	6. 35 - 14 =	7. -6 - 4 =	6	
			7	
			8	
8. -64 - -53 =	9. -49 - 54 =	10. -63 - -78 =	9	
			10	
			Score	

Problem Solving

If eggs cost $1.29 per dozen, how much will 7 dozen cost?

Review Exercises

Notes

1. -6 - 9 =   2. -6 - -9 =

3. 16 - -18 =   4. -66 - 42 =

Helpful Hints	Use what you've learned to solve the problems on this page.	**Examples:** $-7 + 4 + -3 + 2 =$ $-10 + 6 = -$ (sign is negative)	$\begin{array}{r} 10 \\ -\ 6 \\ \hline 4 = \boxed{-4} \end{array}$	$-7 - -6 =$ $-7 + 6 = -$ (sign is negative)	$\begin{array}{r} 7 \\ -\ 6 \\ \hline 1 = \boxed{-1} \end{array}$	$15 - 36 =$ $15 + -36 = -$ (sign is negative)	$\begin{array}{r} 36 \\ -\ 15 \\ \hline 21 = \boxed{-21} \end{array}$

S. -76 + 36 =	S. 9 - -6 =	1. -37 + -16 =	1	
			2	
			3	
2. -92 + 103 =	3. -7 - 8 =	4. 6 - -9 =	4	
			5	
5. -7 + 3 + -8 =	6. 14 + -6 + 3 + -8 =	7. 63 - 96 =	6	
			7	
8. 3 - -12 =	9. -326 + 427 =	10. -273 - 408 =	8	
			9	
			10	
			Score	

Problem Solving	A business needs 325 postcards to mail to customers. If postcards come in packages of 25, how many packages does the business need to buy?

Review Exercises

Notes

1. $3 \times 1\frac{1}{4} =$

2. $6 \div 1\frac{1}{2} =$

3. $\frac{3}{4} \times 16 =$

4. $\frac{4}{5} \div \frac{1}{10} =$

Helpful Hints

The product of two integers with different signs is negative.
The product of two integers with the same sign is positive ($\bullet$ means multiply).

Examples:

$7 \bullet -16 = -$
(sign is negative)

$\begin{array}{r} 16 \\ \times\ 7 \\ \hline 112 \end{array}$ = $\boxed{-112}$

$-8 \bullet -7 = +$
(sign is positive)

$\begin{array}{r} 8 \\ \times\ 7 \\ \hline 56 \end{array}$ = $\boxed{+56}$

S. $-3 \times -16 =$

S. $-18 \bullet 7 =$

1. $-4 \bullet -17 =$

2. $16 \times -4 =$

3. $-24 \bullet -12 =$

4. $23 \times -16 =$

5. $-23 \bullet 32 =$

6. $7 \times -19 =$

7. $-3 \bullet -7 =$

8. $-19 \times -20 =$

9. $32 \bullet -8 =$

10. $-16 \bullet -12 =$

1	
2	
3	
4	
5	
6	
7	
8	
9	
10	
Score	

Problem Solving

At night the temperature was 37°. By morning it had dropped 47°.
What was the temperature in the morning?

Review Exercises

Notes

1. -6 + 7 + -2 + 6 =   2. 3 - -7 =

3. -3 • -9 =   4. -6 x -42 =

Helpful Hints	When multiplying more than two integers, group them in pairs to simplify. An integer next to a parentheses means to multiply. **Examples:**	2 • -3 (-6) = (2 • -3) (-6) = -6 (-6) = + (sign is positive)	6 x 6 36 = (+36)	-2 • -3 • 4 • -2 = (-2 • -3) • (4 • -2) = 6 • -8 = - (sign is negative)	8 x 6 48 = (-48)

S. -3 • 7 • -2 =   S. -3 (6) • -3 =   1. 2 (-3) • 4 =

2. -4 • -3 (-4) =   3. 2 • -3 • -4 • 5 =   4. 6 (3) • -4 x (-5) =

5. 1 • -1 • -3 • -2 =   6. (-2) (-3) (-4) =   7. -8 (-1) • 1 (-4) =

8. 4 (-3) • 2 (-3) =   9. (-3) (-2) (3) (4) =   10. 10 (-11) (-3) =

1	
2	
3	
4	
5	
6	
7	
8	
9	
10	
Score	

Problem Solving

How much will a worker earn in 15 hours if he earns $1\frac{1}{2}$ dollars per hour?

103

Review Exercises

Notes

1. $6\overline{)607}$ 2.   -3 (4) • -5 =

3.   12.3   4.   7 + .63 + 7.18 =
         x 7

Helpful Hints	The quotient of two integers with different signs is negative. The quotient of two integers with the same signs is positive. (HINT: Determine the sign, then divide.)	**Examples:** $36 \div -4 = -$ (sign is negative) $\quad 4\overline{)\begin{array}{r}9\\36\\-36\\\hline 0\end{array}} = (-9)$	$\dfrac{-123}{-3} = +$ (sign is positive) $\quad 3\overline{)\begin{array}{r}41\\123\\-12\downarrow\\\hline 3\end{array}} = (+41)$

			No.	
S. 9 ÷ -3 =	S. $\dfrac{-90}{-15}$ =	1. -64 ÷ 4 =	1	
			2	
			3	
2. -336 ÷ -7 =	3. $\dfrac{-75}{-5}$ =	4. 104 ÷ -4 =	4	
			5	
5. $\dfrac{-110}{-5}$ =	6. 288 ÷ -12 =	7. 42 ÷ -7 =	6	
			7	
8. 714 ÷ -21 =	9. $\dfrac{-65}{-5}$ =	10. 684 ÷ -36 =	8	
			9	
			10	
			Score	

Problem Solving	If the temperature at midnight was -7° and by 3:00 A.M. the temperature had dropped another 19°F, what was the temperature at 3:00 A.M.?

Review Exercises

Notes

1. $-36 \div 4 =$

2. $\dfrac{-56}{-7} =$

3. $3 \cdot 6 \cdot -5 =$

4. $-2 \, (-3) \, (-4) =$

Helpful Hints	Use what you have learned to solve problems like these.	**Examples:**	$\dfrac{-36 \div -9}{4 \div -2} = \dfrac{4}{-2} = \boxed{-2}$ (sign is negative)	$\dfrac{4 \times -8}{-8 \div 2} = \dfrac{-32}{4} = \boxed{+8}$ (sign is positive)

S. $\dfrac{-10 \div 5}{2 \div -1} =$

S. $\dfrac{-4 \cdot -6}{-8 \div 4} =$

1. $\dfrac{-32 \div -4}{-12 \div 3} =$

2. $\dfrac{-6 \times 5}{-30 \div 3} =$

3. $\dfrac{12 \cdot -2}{18 \div 3} =$

4. $\dfrac{-6 \cdot -6}{2 \div -2} =$

5. $\dfrac{54 \div -9}{-18 \div -9} =$

6. $\dfrac{16 \div -2}{-1 \times 4} =$

7. $\dfrac{-75 \div -25}{-3 \div -1} =$

8. $\dfrac{42 \div -2}{-3 \cdot -7} =$

9. $\dfrac{45 \div -5}{-9 \div 3} =$

10. $\dfrac{-56 \div -7}{-36 \div -9} =$

1	
2	
3	
4	
5	
6	
7	
8	
9	
10	
Score	

Problem Solving	A boy scored 627 points on a video game. This was 129 points more than his brother. How many points did his brother score?

105

Solve each of the following.

1. -9 + 7 =   2. 9 + -7 =   3. -9 + -7 =

4. -7 + -8 + 14 =   5. -32 + 16 + 21 + -24 =

6. 7 - 9 =   7. 4 - -9 =   8. -3 - 9 =

9. -13 - 14 =   10. 16 - 17 =   11. 3 • -16 =

12. -4 • -19 =   13. 2 (-7) (-4) =   14. -2 • 3 (-4) • 2 =

15. -36 ÷ 4 =   16. -126 ÷ -3 =   17. $\dfrac{-128}{-8}$ =

18. $\dfrac{-36 \div 2}{24 \div -4}$ =   19. $\dfrac{6 \cdot -3}{-54 \div -6}$ =   20. $\dfrac{-20 \cdot -3}{-30 \div -10}$ =

1	
2	
3	
4	
5	
6	
7	
8	
9	
10	
11	
12	
13	
14	
15	
16	
17	
18	
19	
20	

Review Exercises

Notes

1. Find the area.

7 ft.

2. Find 15% of 65.

3. $\dfrac{3}{5} \times 2\dfrac{1}{2} =$

4. $3\dfrac{1}{2} \div 2 =$

Helpful Hints

Bar graphs are used to compare information.

1. Read the title.
2. Understand the meaning of the numbers. Estimate, if necessary.
3. Study the data.
4. Answer the questions, showing work if necessary.

Use the information in the graph to answer the questions.

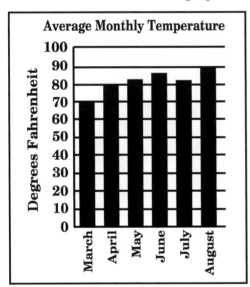

Average Monthly Temperature

Degrees Fahrenheit

(March, April, May, June, July, August)

S. Which month had the second lowest average temperature?

S. How many degrees cooler was the average temperature in April than in August?

1. In which month was the average temperature 81°?

2. In which month did the average temperature drop from the previous month?

3. Which month had the second highest average temperature?

4. How much warmer was the average temperature in August than in May?

5. For what month did the average temperature rise the most from the previous month?

6. Which months had average temperatures less than July's average temperature?

7. Which two month's average temperatures were the closest?

8. The coolest day in August was 77°. How much less than the average temperature was this?

9. What was the increase in average temperature from May to June?

10. Which months had an average temperature less than May's?

1	
2	
3	
4	
5	
6	
7	
8	
9	
10	
Score	

Problem Solving

A fence around a garden is in the shape of a circle. If its diameter is 12 yards, what is the distance around the fence?

Review Exercises

Notes

1. Change $\frac{15}{20}$ to percent.

2. Change .7 to a percent.

3. Find the area.

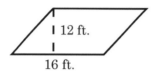

    12 ft.

    16 ft.

4. $12 \overline{\smash{)}2643}$

Use the information in the graph to answer the questions.

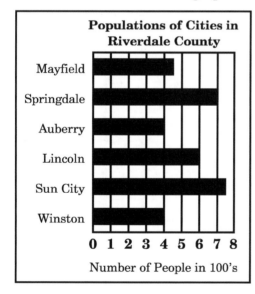

Populations of Cities in Riverdale County

Mayfield
Springdale
Auberry
Lincoln
Sun City
Winston

0 1 2 3 4 5 6 7 8

Number of People in 100's

S. Which two cities had the same population?

S. What is the combined population of Mayfield and Lincoln?

1. Within 3 years, the population of Lincoln is expected to double. What will its population be in 3 years?

2. How many more people live in Springdale than in Auberry?

3. What is the difference in population between the largest and smallest cities?

4. What is the total population of Riverdale County?

5. How many more people live in Sun City than in Mayfield?

6. To reach a population of 900, by how much must Mayfield grow?

7. What is the total population of the two largest cities?

8. How many people must move to Winston before its population is equal to that of Springdale?

9. Which city has nearly double the population of Auberry?

10. What is the total population of all cities whose population is less than 500?

1	
2	
3	
4	
5	
6	
7	
8	
9	
10	
Score	

Problem Solving

A factory can make a part in $1\frac{1}{2}$ minutes. How many parts can be made in 30 minutes?

Review Exercises

Notes

1. Find the area.

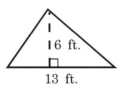

    6 ft.
    13 ft.

2. Classify the triangle.

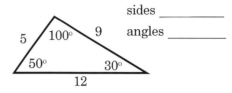

    sides _____
    angles _____

3. 1,000 x 2.365

4.   7.23
      x 6

Helpful Hints

Line graphs are used to show changes and relationships between quantities.

1. Read the title.
2. Understand the meaning of the numbers. Estimate, if necessary.
3. Study the data.
4. Answer the questions, showing work if necessary.

Use the information in the graph to answer the questions.

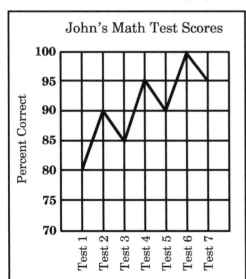

John's Math Test Scores

S. What was John's score on Test 5?

S. How much better was John's score on Test 7 than on Test 3?

1. On which three tests did John score the highest?

2. What is the difference between his highest and lowest score?

3. Find John's average score by adding all his scores and dividing by the number of scores.

4. How many scores were below John's average score?

5. What are his two lowest scores?

6. What is the average of his highest and lowest score?

7. How much higher was Test 4 than Test 1?

8. What was the difference between his highest score and his second lowest score?

9. How many test scores were improvements over the previous test?

10. Did John's progress generally improve or get worse?

1	
2	
3	
4	
5	
6	
7	
8	
9	
10	
Score	

Problem Solving

Movie passes cost $3.75 each. How much will it cost for 7 passes?

109

Review Exercises

Notes

1. $\frac{4}{5}$
    $+ \frac{3}{5}$
    _____

2. $\frac{7}{8}$
    $- \frac{1}{8}$
    _____

3. 3
    $- 1\frac{1}{7}$
    _____

4. $3\frac{1}{3}$
    $- 1\frac{2}{3}$
    _____

Helpful Hints

1. Read the title.
2. Understand the meaning of the numbers. Estimate, if necessary.
3. Study the data.
4. Answer the questions, showing work if necessary.

Use the information in the graph to answer the questions.

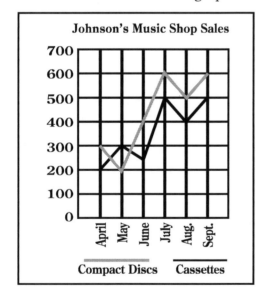

Johnson's Music Shop Sales

Compact Discs   Cassettes

S. In which months were the most cassettes sold?

S. How many more compact discs than cassettes were sold in August?

1. What was the total amount of compact discs sold in May and August?

2. How many more compact discs were sold in July than in August?

3. In which month were more cassettes sold than compact discs?

4. During which month did compact discs outsell cassettes by the most?

5. In September, how many more compact discs were sold than cassettes?

6. Which two months had the highest total sales?

7. What was the total number of compact discs and cassettes sold in September?

8. What was the increase in sales of compact discs between May and June?

9. What was the decrease in sales of compact discs between July and August?

10. What were the two lowest total sales months?

1	
2	
3	
4	
5	
6	
7	
8	
9	
10	
Score	

Problem Solving

Bill's test scores were 75, 80 and 100. What was his average score?

Review Exercises

Notes

1. 32 + -76 =

2. 6 - -9 =

3. 3 • -7 =

4. Find 12% of 250.

Helpful Hints

A circle graph shows the relationship between the parts to the whole and to each other.

1. Read the title.
2. Understand the meaning of the numbers. Estimate, if necessary.
3. Study the data.
4. Answer the questions, showing work if necessary.

Use the information in the graph to answer the questions.

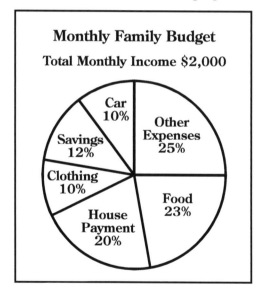

Monthly Family Budget

Total Monthly Income $2,000

S. What percent of the family budget is spent for food?

S. After the car payment and house payment are paid, what percent of the budget is left?

1. What percent of the budget is used to pay for food and clothing?

2. How much is spent on food each month? (Hint: Find 23% of $2,000.)

3. How many dollars are spent on clothing?

4. How many dollars was the house payment?

5. What percent of the budget did the three largest items represent?

6. What percent of the budget is for savings, clothing, and the car payment?

7. In twelve months, what is the total income?

8. What percent of the budget is left after the house payment has been made?

9. Which two items require the same part of the budget?

10. Which part of the budget would pay for medical expenses?

1	
2	
3	
4	
5	
6	
7	
8	
9	
10	
Score	

Problem Solving

If a plane can travel 350 miles per hour, how far can it travel in 3.5 hours?

Review Exercises

Notes

1. Reduce $\dfrac{12}{16}$ to its lowest terms.

2. Change $\dfrac{6}{8}$ to a percent.

3. 15 is what % of 20?

4. -33 + 16 =

Helpful Hints

Circle graphs can be used to show fractional parts.

1. Read the title.
2. Understand the meaning of the numbers.
3. Study the data.
4. Answer the questions.

Use the information in the graph to answer the questions.

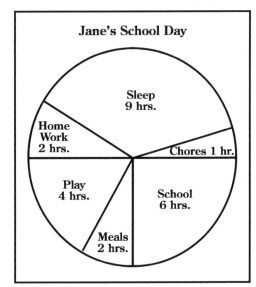

Jane's School Day

Sleep 9 hrs.
Home Work 2 hrs.
Chores 1 hr.
Play 4 hrs.
School 6 hrs.
Meals 2 hrs.

S. What fraction of the day does Jane play?

S. What fraction of Jane's day is used for school?

1. How many more hours does Jane sleep per day than play?

2. How many hours of homework does Jane have in a week?

3. How many hours per day are school- related activities?

4. What fraction of the day is spent for school, homework, and chores?

5. What fraction of the day does Jane spend for school, sleep, and chores?

6. How many hours does Jane spend in school in 3 weeks?

7. If Jane goes to bed at 9:00 P.M., what time does she get up in the morning?

8. If school starts at 8:30 A.M., what time is school dismissed?

9. How many hours per week does Jane spend in school and on homework?

10. What fractional part of the day does Jane play?

1	
2	
3	
4	
5	
6	
7	
8	
9	
10	
Score	

Problem Solving

A bedroom is in the shape of a rectangle with length 12 feet and width 14 feet. How many square feet of wall-to-wall carpet would it take to cover the floor?

Review Exercises

Notes

1. 7106 - 774 =   2. 76 x 403 =

3. 667 + 19 + 246 =   4. 5 ⟌ 5015

Helpful Hints	Picture graphs are another way to compare statistics.	1. Read the title. 2. Understand the meaning of the numbers. Estimate, if necessary. 3. Study the data. 4. Answer the questions.

Use the information in the graph to answer the questions.

Bikes Made By Street Bike Company

1986 🚲 🚲 🚲
1987 🚲 🚲 🚲 🚲
1988 🚲 🚲 🚲 🚲
1989 🚲 🚲 🚲 🚲 🚲 🚲
1990 🚲 🚲 🚲 🚲 🚲
1991 🚲 🚲 🚲 🚲 🚲 🚲 🚲 🚲

Each 🚲 represents 1,000 bikes

S. How many bikes were made in 1989?

S. How many more bikes were made in 1991 than in 1988?

1. Which year produced twice as many bikes as 1986?

2. What was the total number of bikes that the company produced in 1990 and 1991?

3. Which two years did the company make the most bikes?

4. 1992 is reported to be double the production of 1988. How many bikes are to be produced in 1992?

5. What is the total number of bikes produced in 1986 and 1991?

6. It cost $50 to make a bike in 1986, how much did the company spend making bikes that year?

7. The cost to make a bike jumped to $75 in 1991. How much did the company spend making bikes in 1991?

8. What is the difference in bikes produced in 1986 and 1991?

9. What is the total number of bikes made during the company's 3 most productive years?

10. One half of the bikes made in 1989 were ladies' style. How many ladies' bikes were made in 1989?

1	
2	
3	
4	
5	
6	
7	
8	
9	
10	
Score	

Problem Solving	Bill, John, Mary, and Sheila together earned $524. If they wanted to share the money equally, how much would each one receive?

Review Exercises

Notes

1. Find the perimeter of a rectangle with length 19 inches and width 13 inches.

2. Find the circumference of a circle with radius 3 feet.

3. Find the area of a rectangle with length 26 inches and width 15 inches.

4. Find the area of a triangle with base 8 feet and height 11 feet.

Helpful Hints

1. Read the title.
2. Understand the meaning of the symbols. Estimate, if necessary.
3. Study the data.
4. Answer the questions.

Use the information in the graph to answer the questions.

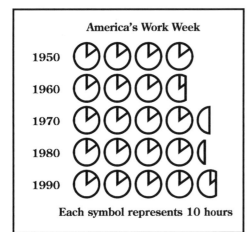

America's Work Week

1950
1960
1970
1980
1990

Each symbol represents 10 hours

S. Which work week was the longest?

S. How many hours shorter was the work week in 1960 than in 1990?

1. How many hours long was the workweek in 1970?

2. How may hours did the work week increase between 1980 and 1990?

3. If the average employee works 50 weeks per year how many hours did he work in 1950?

4. Which year's work week was approximately 38 hours?

5. Which years had the 2 shortest work weeks?

6. How many hours less was the work week in 1950 than 1980?

7. If the work week is 5 days, what was the average number of hours worked per day in 1950?

8. In 1990, if an employee decided to work 4 days per week, what would his average number of hours be per day?

9. Any work time over 40 hours is overtime. What was the average worker's weekly overtime in 1970?

10. What is the difference between the longest and the shortest work week?

1	
2	
3	
4	
5	
6	
7	
8	
9	
10	
Score	

Problem Solving

A man earned $5\frac{1}{4}$ dollars per hour. How much would he earn in 5 hours?

Height of Waterfalls

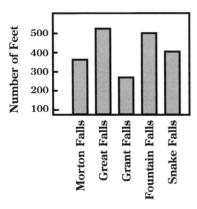

1. Which waterfall is the smallest?
2. Approximately how high is Great Falls?
3. Approximately how much higher is Morton Falls than Grant Falls?
4. Which waterfall is about the same height as Morton Falls?
5. Which waterfall is the fourth highest?

Family Budget: $3,000 per month

6. What percent of the family's money is spent on clothing?
7. What percent of the budget is spent on housing?
8. How many dollars are spent on housing per month? (Hint: Find 20% of $3,000)
9. How many dollars do they save per month?
10. What percent of the family budget is left after paying food, housing and car expenses?

Average Monthly Temperatures

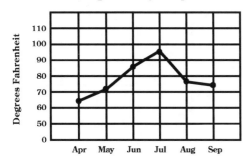

11. What is the average temperature of May?
12. How much cooler was April than July?
13. Which was the second hottest month?
14. Which month's temperature dropped the most from the previous month?
15. What is the difference in temperature between the hottest month and the second hottest month?

Fish caught in Drakes Bay in 1991

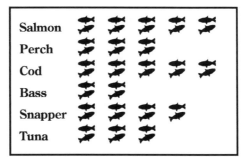

Each symbol represents 10,000 fish

16. How many perch were caught in 1991?
17. How many more snapper were caught than bass?
18. How many salmon and cod were caught?
19. If the average perch weighs 3 pounds, how many pounds were caught in 1991?
20. What are the three most commonly caught types of fish?

1	
2	
3	
4	
5	
6	
7	
8	
9	
10	
11	
12	
13	
14	
15	
16	
17	
18	
19	
20	

Review Exercises

Notes

1. $376 + 92 + 743 =$ 2. $2,106 - 1,567 =$

3. $\begin{array}{r} 724 \\ \times\ 16 \\ \hline \end{array}$ 4. $7\overline{)1137}$

Helpful Hints

1. Read the problem carefully.
2. Find the important facts and numbers.
3. Decide what operations to use.
4. Solve the problem.

* Sometimes drawing a practice diagram can help.
* Sometimes a formula is necessary.
* Sometimes reading a problem more than once helps.
* Show your work and put answers in a phrase.

S. There are three sixth grade classes with enrollments of 36, 37, and 33. How many sixth graders are there in all?

S. A family drove 355 miles each day for 7 days. How far did the family drive altogether?

1. The attendance at the Hawk's game last year was 38,653. This year 45,629 attended. What was the increase in attendance this year?

2. 7 friends earned 1,463 dollars. If they want to divide the money evenly, how much will each person receive?

3. A school put on a play. If 467 attended the Tuesday performance, 655 attended the Saturday performance, and 596 attended the Sunday performance, what was the total attendance?

4. In an election, John received 2,637 votes. Julie received 2,904 votes. How many more votes did Julie receive than John?

5. If a plane travels at 750 miles per hour, how far does it travel in 12 hours?

6. A car traveled 295 miles in five hours. What was its average speed?

7. How far can a car go if the tank contains 12 gallons of gas and it can travel 23 miles per gallon?

8. A field is in the shape of a triangle with sides 265 feet, 379 feet and 189 feet. How many feet is it around the field?

9. A rope is 544 feet long. If it is cut into pieces 2 feet long, how many pieces will there be?

10. A library has 2,365 fiction books, 2,011 nonfiction books, and 796 reference books. How many books does the library have in all?

1	
2	
3	
4	
5	
6	
7	
8	
9	
10	
Score	

Review Exercises

Notes

1. $\dfrac{3}{5}$
   $+\dfrac{1}{2}$
   _____

2. $\dfrac{7}{8}$
   $-\dfrac{1}{4}$
   _____

3. $2\dfrac{1}{2} \times 3 =$

4. $7\dfrac{1}{2} \div 1\dfrac{1}{2} =$

Helpful Hints

1. Read the problem carefully.
2. Find the important facts and numbers.
3. Decide what operations to use.
4. Solve the problem.

* Sometimes drawing a practice diagram can help.
* Sometimes a formula is necessary.
* Sometimes reading a problem more than once helps.
* Show your work.

S. A baker uses $2\dfrac{1}{4}$ cups of flour for a cake and $1\dfrac{3}{5}$ cups of flour for a pie. How much flour did he use?

S. A ribbon is $5\dfrac{1}{2}$ long. It was cut into pieces $\dfrac{1}{2}$ foot long. How many pieces were there?

1. Bill weighed $124\dfrac{1}{4}$ pounds three months ago. If he now weighs $132\dfrac{1}{2}$ pounds, how many pounds did he gain?

2. Steve earned 60 dollars and spent $\dfrac{2}{3}$ of it. How much did he spend?

3. It is $2\dfrac{1}{2}$ miles around a race track. How far will a car travel in 12 laps?

4. It takes a man $12\dfrac{1}{2}$ minutes to drive to work and $16\dfrac{3}{4}$ minutes to drive home. What is his total commute time?

5. What is the perimeter of a square flower bed which is $8\dfrac{1}{2}$ feet on each side?

6. Andy brought 6 melons, the total weight of which were 16 pounds. What was the average weight of each melon? (Express your answer as a mixed number.)

7. If a cook had $5\dfrac{1}{4}$ pounds of beef and used $3\dfrac{3}{4}$ pounds, how much beef was left?

8. Sue worked $7\dfrac{1}{2}$ hours on Monday and $6\dfrac{1}{4}$ hours on Tuesday. How many hours did she work in all?

9. A car can travel 50 miles per hour. At this rate, how far will it travel in $2\dfrac{1}{2}$ hours?

10. A factory can produce a tire in $2\dfrac{1}{2}$ minutes. How many tires can it produce in 40 minutes?

1	
2	
3	
4	
5	
6	
7	
8	
9	
10	
Score	

Review Exercises

Notes

1. $3.26 + $4.19 + $6.24 =

2.
$$\begin{array}{r} \$7.00 \\ -\ \$3.56 \\ \hline \end{array}$$

3.
$$\begin{array}{r} \$3.92 \\ \times\ \ 6 \\ \hline \end{array}$$

4. $7\overline{)\ \$15.33}$

Helpful Hints

1. Read the problem carefully.
2. Find the important facts and numbers.
3. Decide what operations to use
4. Solve the problem.

* Sometimes drawing a practice diagram can help.
* Sometimes a formula is necessary.
* Sometimes reading a problem more than once helps.
* Show your work and put answers in a phrase.

S. If 6 dozen pencils cost $8.64, what is the cost of 1 dozen pencils?

S. Find the average speed of a jet that traveled 1350 miles in 2.5 hours.

1. Potatoes cost $3.19 a pound and carrots cost $2.78 per pound. How much more do potatoes cost per pound?

2. A man bought a chair for $125.50, a desk for $279.25, and a desk lamp for $24.95. What was the total cost of those items?

3. What is the area of a rectangular farm that is 1.8 miles long and 1.3 miles wide?

4. If 12 cans of corn cost $13.68, what is the cost of one can?

5. A plane can travel 240 miles in one hour. At this rate, how far can it travel in .8 hours?

6. A sack of potatoes cost $2.70. If the price was $.45 per pound, how many pounds of potatoes were in the sack?

7. A baseball mitt was on sale for $24.95. If the regular price was $35.50, how much could be saved by buying it on sale?

8. Tom weighs 129.6 pounds and Bill weighs 135.25 pounds. What is their total weight?

9. If 6 pounds of butter costs $5.34, what is the price per pound?

10. An engine uses 2.5 gallons of gas per hour. How many gallons will it use in 3.2 hours?

1	
2	
3	
4	
5	
6	
7	
8	
9	
10	
Score	

Review Exercises

Notes

1. Find 16% of 230

2. What is 20% of 25?

3. Find the area

7 ft.

4. Find the area

16 ft.

9 ft.

| **Helpful Hints** | Use what you have learned to solve the following problems. | 1. Read the problem carefully.
2. Find the important facts and numbers.
3. Decide what operations to use
4. Solve the problem. |

S.   A rope 6 feet long is to be cut into pieces $1\frac{1}{2}$ feet long. How many pieces will there be?

S.   If cans of soda cost $.35, how many cans of soda can be bought with $5.60?

1.   What is the perimeter of a square field which has sides 139 feet long?

2.   Steak cost $4.80 per pound. How much will .7 pounds cast?

3.   Sam had a total score of 356 points on 4 tests. What was his average score?

4.   A carpenter needs to glue together three boards with thickness of 3.9 inches, 2.25 inches, and 1.875 inches. What would the total thickness be after they are glued together?

5.   A man earns $3\frac{1}{2}$ dollars per hour. How much will he earn in 8 hours?

6.   A board was 9 feet long. If $2\frac{3}{4}$ feet was cut off, how much of the board was left?

7.   After shopping for groceries a man had $13.64. He started with $50.00, how much did the groceries cost?

8.   At a junior high school there are 600 seventh graders. If they are to be grouped into 15 equally sized homerooms, how many will be in each homeroom?

9.   Sun City's population is 202,516. Elk Grove's population is 178,319. How much larger is Sun City's population?

10. A rancher owns 360 cows. If he decides to sell $\frac{3}{4}$ of them, how many will he sell?

1	
2	
3	
4	
5	
6	
7	
8	
9	
10	
Score	

Review Exercises

Notes

1. $2\overline{)73}$

2. $5\overline{)1172}$

3. $12\overline{)763}$

4. $25\overline{)5265}$

Helpful Hints	1. Read the problem carefully.
	2. Find the important facts and numbers.
	3. Decide what operations to use and in what order to do them.
	4. Solve the problem.

S.   Bill's test scores were 78, 87 and 96. What was his average score?

S.   A store owner bought seven crates of lettuce that weighed 120 lbs. each. He also bought 12 sacks of potatoes that weighed a total of 125 lbs. What was the total weight of the lettuce and potatoes?

1.   A man decides to buy a car. He makes a down payment of $500 and agrees to pay 36 monthly payment of $250. How much will the car cost him?

2.   Last week a man worked 9 hours per day for five days. This week he worked 36 hours. How many hours did he work in all?

3.   A farmer has an orchard with 36 rows of trees. There are 22 trees in each row. If each tree produces 14 bushels of fruit, how many bushels of fruit will be produced in all?

4.   A junior high school has 7 eighth grade classes with 36 students in each class. If 506 students attend the school, how many are not in the eighth grade?

5.   Buses hold 75 people. If 387 student and 138 parents are attending a football game, how many buses will they need?

6.   A seventh grade has 314 boys and 310 girls. If they are to be grouped into equal classes of 26 each, how many classes will there be?

7.   A car traveled 340 miles each day for 5 days and 250 miles on the sixth day. How many miles did it travel in all?

8.   A tank holds 200,000 gallons of fuel. If 57,000 gallons were removed one day and 62,000 gallons the next, how many gallons are left?

9.   Nine buses hold 85 people each. If 693 people buy tickets for a trip, how many seats won't be taken?

10.   Jim worked 50 weeks last year. Each week he worked 36 hours. If he worked an additional 220 hours overtime, how many hours did he work in all last year?

1	
2	
3	
4	
5	
6	
7	
8	
9	
10	
Score	

Review Exercises

Notes

1. Find the average of 24, 42, and 45

2. Find $\frac{1}{2}$ of $3\frac{1}{2}$

3. $2\frac{6}{8} + 1\frac{1}{3} =$

4.
$$2\frac{3}{4}$$
$$+\ 3\frac{1}{4}$$
$$\overline{}$$

Helpful Hints

1. Read the problem carefully.
2. Find the important facts and numbers.
3. Decide what operations to use.
4. Solve the problem.

* Sometimes drawing a practice diagram can help.
* Sometimes a formula is necessary.
* Sometimes reading a problem more than once helps.
* Show your work and put answers in a phrase.

S. A tailor had $8\frac{1}{2}$ yards of cloth. He cut off 3 pieces which were $1\frac{1}{2}$ yards each. How much of the cloth was left?

S. A man had 56 dollars. If he gave $\frac{1}{2}$ of it to his son, how much did he have left?

1. A painter needs 7 gallons of paint. He already has $2\frac{1}{2}$ gallons in one bucket and $3\frac{1}{4}$ gallons in another. How many more gallons does he need?

2. There are 30 people in a class. If $\frac{2}{5}$ of them are boys then how many are girls?

3. Bill worked $5\frac{1}{2}$ hours on Monday and $6\frac{3}{4}$ hours on Tuesday. If he was paid 8 dollars per hour, what was his pay?

4. Susan made 36 bracelets last week and 42 this week. If she sold half of them, how many bracelets did she sell?

5. Joe's ranch has 4,000 acres. $\frac{1}{4}$ of the ranch was for crops. $\frac{2}{3}$ of the remainder was used for grazing. How many acres were for grazing?

6. A family was going to take a 400 mile trip. If they traveled $\frac{1}{4}$ of the distance the first day, how many miles were left to travel?

7. A store has 20 feet of copper wire. If the wire is cut into pieces $2\frac{1}{2}$ feet long and each piece sells for 6 dollars, how much would all the pieces cost?

8. A garden in the shape of a rectangle is 30 feet long and 20 feet wide. If $\frac{1}{3}$ of it is to be used for onions, how many square feet of the garden will be used for onions?

9. A farmer picked $6\frac{1}{2}$ bushels of fruit each day for 5 days. He then sold $15\frac{1}{2}$ bushels. How many bushels were left to sell?

10. A man bought $12\frac{1}{2}$ pounds of beef. He put $10\frac{1}{4}$ lbs. in his freezer. He used $\frac{3}{5}$ of the rest for cooking a stew. How many pounds did he use for the stew?

1	
2	
3	
4	
5	
6	
7	
8	
9	
10	
Score	

Review Exercises

Notes

1. $4.3 + 7.6 + 7.97 =$ 2. $\begin{array}{r} .63 \\ \times\ 3.5 \\ \hline \end{array}$

1. $2.013 - 1.66 =$ 4. $3\,\overline{\smash{)}\,\$6.30}$

Helpful Hints

1. Read the problem carefully.
2. Find the important facts and numbers.
3. Decide what operations to use.
4. Solve the problem.

* Sometimes drawing a practice diagram can help.
* Sometimes a formula is necessary.
* Be careful about decimal placement.
* Show your work and put answers in a phrase.

S.   A man bought 5 bags of chips at $1.39 each, and a large pizza for $9.95. How much did he spend?

S.   Mrs. Jones bought a hammer for $6.75 and a screwdriver for $5.19. If she gave the clerk a $20 bill, how much change would she receive?

1.   A man bought a car. He paid $1500.00 down and $250 per month for 36 months. How much did he pay altogether?

2.   Jill is taking a trip of 126 miles. If her car gets 21 miles per gallon and gas cost $1.19 per gallon, how much will the trip cost?

3.   Susan worked 30 hours and was paid $4.15 per hour. If she bought a pair of shoes for $16.55, how much of her pay was left?

4.   Cans of peas are on sale at 2 for $1.19. How much would 12 cans cost?

5.   Jeans are on sale for $9.95 a pair. If the regular price is $12.25, how much would be saved by buying 3 pairs of jeans on sale?

6.   Three friends earned $3.65 on Monday, $4.75 on Tuesday, and $3.75 on Wednesday. If they divided the money evenly, how much would each receive?

7.   A jacket cost $12. If the sales tax was 7%, what was the total price of the jacket?

8.   A man bought 12 gallons of gas at $1.16 per gallon and a quart of oil for $3.19. How much did he spend altogether?

9.   A floor is 12 by 10 feet. If carpet costs $7.00 per square foot, how much will it cost to carpet the floor completely?

10.   Cans of peas are 3 for $1.29. How much would one can of peas and a bag of chips priced at $2.19 cost altogether?

1	
2	
3	
4	
5	
6	
7	
8	
9	
10	
Score	

Review Exercises

Notes

1. Find 20% of 72

2. 6 is what percent of 24?

3. Find the circumference

6 ft.

4. Find the perimeter

26 inches

14 inches

Helpful Hints | Use what you have learned to solve the following problems.

S. A theater has 12 rows of 8 seats. If $\frac{2}{3}$ of them are taken, how many are taken?

S. Tom's times for the 100 yard dash were 11.8, 12.2 and 12.3 What was his average time?

1. John bought a shirt for $13.55 and a pair of shoes for $27.50. If he gave the clerk a $50 bill, how much change would he receive?

2. A group of hikers set out on a 50 mile trip. If they hiked 9 miles per day for three days, how many miles were left to hike?

3. Each of 3 classes has 32 students. If $\frac{2}{3}$ of these students are boys, how many boys are there?

4. If 2 pounds of chicken cost $3.60, how much will 10 pounds cost?

5. Mary is buying a bike. She makes a down payment of $75.00 and pays the rest in 12 monthly installments of $32.00 each. How much does she pay in all?

6. An electrician has 12 feet of wire. If he cuts it into pieces $1\frac{1}{2}$ feet long an sells each piece for $3.00, how many pieces are there and how much will they cost altogether?

7. Bill worked 35 hours at $5.00 per hour. If he wants to buy a bike for $225.00, how much more must he earn?

8. When a man started on a diet he weighed $160\frac{1}{2}$ pounds. The first week he lost $3\frac{1}{4}$ pounds and the second week he lost $2\frac{1}{2}$ pounds. How much did he weigh after 2 weeks?

9. A man bought 3 pens for $1.19 each, and a notebook for $3.15. How much did he spend altogether?

10. A car traveled 96 miles. If it averaged 12 miles per gallon of gas and gas cost $1.29, how much did the trip cost?

1	
2	
3	
4	
5	
6	
7	
8	
9	
10	
Score	

Review Exercises

Notes

1. Classify by sides   2. Classify by angles

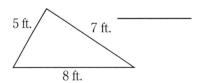

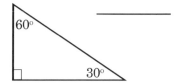

3. Find the perimeter of an equilateral triangle with sides of 39 feet.

4. Find the area of a square with sides of 16 feet.

Helpful Hints

1. Read the problem carefully.
2. Find the important number and facts.
3. Decide which operations to use and in what order
4. Solve the problem.

Re-read the problems.
Draw a diagram or picture if necessary.
Use a formula if necessary.

S. Bill bought 6 dozen hot dogs at $2.29 a dozen and 3 dozen burger patties at $6.15 a dozen. How much did he spend altogether?

S. A rectangular yard is 16 feet by 24 feet. How many feet of fencing is required to enclose it? If each 5-foot section costs $35.00, how much will the fencing cost?

1. Two trains leave a station in opposite directions, one at 85 miles per hour, the other at 75 miles per hour. How far apart will they be after 3 hours?

2. A factory can make a table in 320 minutes and a chair in 8 minutes. How long will it take to make 7 tables and 15 chairs? Express the answer in hours and minutes.

3. Mr. Jones' class has 32 students and Mrs. Jensen's class has 40 students. 25% of Mr. Jones' class got A's and 15% of Mrs. Jensen's class got A's. How many students got A's?

4. John can buy a car by paying $3,000 down and making 36 monthly payment of $150 or by paying nothing down and making 48 payments of $180. How much would he save by paying the first way?

5. There are 400 boys and 350 girls at Anderson School. $\frac{3}{4}$ of the boys take the bus and $\frac{3}{5}$ of the girls take the bus. How many students in all take the bus?

6. A farmer had 600 pounds of apples. He gave 200 pounds to neighbors and sold $\frac{1}{2}$ of the remainder at $.75 per pound. How much did he make selling the apples?

7. Tom finished a 30 mile walkathon. He collected pledges of $26 for each of the first 25 miles and $36 for each remaining mile. How much did he collect in all?

8. A developer has 3 plats of land. One was 39 acres, another was 37 acres, and another of 49 acres. How many $\frac{1}{4}$ acre lots did he have? If each lot sold for $2,000.00 how much did he sell all the lots for?

9. A farm has 2000 acres. If $\frac{3}{5}$ of this was used for crops and $\frac{3}{4}$ of the remainder was used for grazing, how many acres were left?

10. A carpenter bought 3 hammers for $7.99 each, 2 saws for $12.15 each and a drill for $26.55. How much did he spend in all?

1	
2	
3	
4	
5	
6	
7	
8	
9	
10	
Score	

1	
2	
3	
4	
5	
6	
7	
8	
9	
10	
11	
12	
13	
14	
15	
16	
17	
18	
19	
20	

1. Lincoln High School's enrollment is 5,679 and Jefferson High School's enrollment is 4,968. What is the total enrollment for both schools?

2. If a car travels 480 miles in 8 hours, what is its average speed?

3. A class has 35 students. If $\frac{2}{5}$ of them are boys, how many boys are in the class?

4. A rope $7\frac{1}{2}$ feet long was cut into $1\frac{1}{2}$ foot pieces. How many pieces were there?

5. Tim worked $12\frac{1}{4}$ hours on Monday and $9\frac{2}{3}$ hours on Tuesday. How much longer did he work on Monday?

6. A plane can travel 360 miles in one hour. How far can it go in .4 hours at this rate?

7. 6 pounds of corn costs $4.14. What is the price per pound?

8. A shirt was on sale at $12.19. If the regular price was $15.65, how much is saved by buying it on sale?

9. Bill's test scores were 78, 63 and 96. What was his average score?

10. A car traveled 265 miles each day for 7 days and 325 days on the eighth day. How many miles did it travel in all?

11. A tailor has 7 yards of cloth. He cut 2 pieces which were $1\frac{1}{3}$ yards each. How much cloth was left?

12. There are 45 students in a class. If $\frac{3}{5}$ of them are boys, how many are girls?

13. A car traveled 120 miles. It averaged 30 miles per gallon. If gas cost $1.09 per gallon, how much did the trip cost?

14. A boy bought 4 sodas for $.39 each and a bag of chips for $1.19. How much did he spend in all?

15. A rectangular living room floor is 12 feet by 14 feet. If carpet is $2.00 per square foot, how much will it cost to carpet the entire floor?

16. Cans of beans are 3 for $.69. How much would 12 cans cost?

17. A woman bought a car for $6,000.00. If she made a $2,000.00 down payment and paid the rest in equal payments of $400, how many payments would she make?

18. A ranch had 3,000 acres. $\frac{1}{3}$ of the ranch was used for grazing. $\frac{3}{5}$ of the rest was used for crops. How many acres were used for crops?

19. A plumber purchased 3 sinks for $129.00 each and 6 faucets for $16.35. How much did she spend in all?

20. A man wants to put a fence around a square lot with sides of 68 feet. How many feet of fencing is needed? If each 8-foot section of fence costs $25.00, how much will the fence cost?

Solve each of the following.

1.   347   2.   614   3. 6,403 + 763 + 16,799 =
    + 467   723
                17
         + 824

4. 6,502 + 2,134 + 654 + 24 =   5. 6,093 + 748 + 83 + 769 =

6.   927   7.   5,392   8. 6,053 − 4,639 =
   − 648   − 1,764

9. 5,000 − 3,286 =   10. 6,003 − 719 =   11.   73
                                            x 4

12.   7,136   13.   45   14.   342
    x   4   x 37   x 46

15.   643   16. 4 ⟌ 526   17. 4 ⟌ 1376
    x 246

18. 40 ⟌ 568   19. 30 ⟌ 7614   20. 18 ⟌ 1243

1	
2	
3	
4	
5	
6	
7	
8	
9	
10	
11	
12	
13	
14	
15	
16	
17	
18	
19	
20	

Solve each of the following.

1. $$\dfrac{4}{7}$$
   $$+\;\dfrac{1}{7}$$

2. $$\dfrac{7}{8}$$
   $$+\;\dfrac{3}{8}$$

3. $$\dfrac{3}{5}$$
   $$+\;\dfrac{1}{3}$$

4. $$3\,\dfrac{1}{2}$$
   $$+\;2\,\dfrac{3}{8}$$

5. $$7\,\dfrac{3}{5}$$
   $$+\;6\,\dfrac{7}{10}$$

6. $$\dfrac{7}{8}$$
   $$-\;\dfrac{1}{8}$$

7. $$6\,\dfrac{1}{4}$$
   $$-\;2\,\dfrac{3}{4}$$

8. $$5$$
   $$-\;2\,\dfrac{1}{7}$$

9. $$7\,\dfrac{3}{5}$$
   $$-\;2\,\dfrac{1}{2}$$

10. $$7\,\dfrac{1}{4}$$
    $$-\;2\,\dfrac{1}{3}$$

11. $$\dfrac{3}{5} \times \dfrac{2}{7} =$$

12. $$\dfrac{3}{20} \times \dfrac{5}{11} =$$

13. $$\dfrac{5}{6} \times 24 =$$

14. $$\dfrac{5}{8} \times 3\,\dfrac{1}{5} =$$

15. $$2\,\dfrac{1}{2} \times 3\,\dfrac{1}{2} =$$

16. $$\dfrac{5}{6} \div \dfrac{1}{3} =$$

17. $$2\,\dfrac{1}{3} \div \dfrac{1}{2} =$$

18. $$2\,\dfrac{2}{3} \div 2 =$$

19. $$5\,\dfrac{1}{2} \div 1\,\dfrac{1}{2} =$$

20. $$6 \div 1\,\dfrac{1}{3} =$$

1	
2	
3	
4	
5	
6	
7	
8	
9	
10	
11	
12	
13	
14	
15	
16	
17	
18	
19	
20	

Solve each of the following.

1. $\begin{array}{r} 4.67 \\ 3.5 \\ +\ \ 3.743 \\ \hline \end{array}$ 2.   .6 + 7.62 + 6.3 =   3.   16.8 + 6 + 7.9 =

4. $\begin{array}{r} 36.4 \\ -\ 17.8 \\ \hline \end{array}$ 5. $\begin{array}{r} 6.3 \\ -\ 3.69 \\ \hline \end{array}$ 6.   72 − 1.68 =

7. $\begin{array}{r} 2.64 \\ \times\ \ 3 \\ \hline \end{array}$ 8. $\begin{array}{r} 2.6 \\ \times\ 73 \\ \hline \end{array}$ 9. $\begin{array}{r} .63 \\ \times\ 2.4 \\ \hline \end{array}$

10. $\begin{array}{r} .126 \\ \times\ 4.23 \\ \hline \end{array}$ 11.   10 x 3.65 =   12.   1,000 x 3.6 =

13. $2\,\overline{)\,4.64}$ 14. $5\,\overline{)\,6.7}$ 15. $.4\,\overline{)\,.124}$

16. $.004\,\overline{)\,1.2}$ 17. $.15\,\overline{)\,.0045}$ 18. $.67\,\overline{)\,8.71}$

19.   Change $\dfrac{3}{5}$ to a decimal.   20.   Change $\dfrac{7}{20}$ to a decimal.

1	
2	
3	
4	
5	
6	
7	
8	
9	
10	
11	
12	
13	
14	
15	
16	
17	
18	
19	
20	

128

Change numbers 1 through 5 to a percent.

1. $\dfrac{13}{100}$ = 2. $\dfrac{3}{100}$ = 3. $\dfrac{7}{10}$ = 4. .19 = 5. .6 =

Change numbers 6 through 8 to a decimal and a fraction expressed in lowest terms.

6. 8% = .   = —— 7. 18% = .   = —— 8. 80% = .   = ——

Solve the following problems. Label the word problem answers.

9. Find 3% of 74.   10. Find 40% of 320.

11. Find 16% of 400.   12. 4 is what percent of 5?

13. 18 is what % of 24?   14. 20 is what % of 25?

15. Change $\frac{16}{20}$ to a %.   16. Change $\frac{3}{5}$ to a percent.

17. 640 students attend Lincoln School. If 40% of the students are girls, how many girls attend Lincoln School?

18. A team played 40 games. If they won 65% of them, how many games did the team win?

19. A pitcher threw 40 pitches. If 30 were strikes, what percent were strikes?

20. Jim took a test with 20 questions. If he got 18 correct, what percent did he get incorrect?

1	
2	
3	
4	
5	
6	
7	
8	
9	
10	
11	
12	
13	
14	
15	
16	
17	
18	
19	
20	

Use the figure to answer questions 1-8

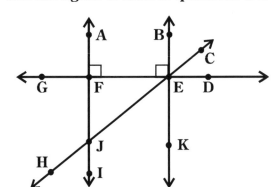

1. Name 2 parallel lines.
2. Name 2 perpendicular lines.
3. Name 4 line segments.
4. Name 4 rays.
5. Name 2 acute angles.
6. Name 2 obtuse angles.
7. Name 1 straight angle.
8. Name 2 right angles.

Triangle A

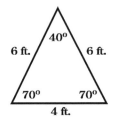

Triangle B

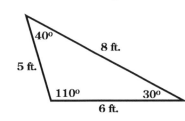

9. Classify triangle A by its sides and angles.

10. Classify triangle B by its sides and angles.

11. Find the perimeter.

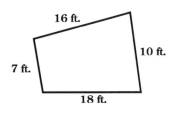

12. Find the circumference.

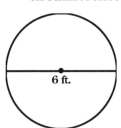

13. Find the area.

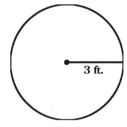

14. Find the area.

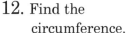
9 ft.
6 ft.

15. Find the area.

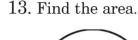
7 ft
14 ft.

16. Find the area.

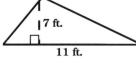
7 ft.
11 ft.

17. Find the area.

14 ft.

18. Identify and count the number of faces, edges, and vertices.

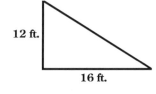

Name: _____   Faces: _____
Edges _____   Vertices: _____

19. Find the area.

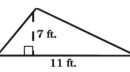
12 ft.
16 ft.

20. Find the perimeter of a square with sides of 37 ft.

1	
2	
3	
4	
5	
6	
7	
8	
9	
10	
11	
12	
13	
14	
15	
16	
17	
18	
19	
20	

1. -8 + 5 =   2. 8 + -5 =   3. -8 + -5 =

4. -6 + 3 + -4 + 2 =   5. -46 + 16 + 23 + -17 =

6. 6 - 9 =   7. 4 - -7 =   8. -3 - 9 =

9. -12 - 15 =   10. 15 - 19 =   11. 4 • -8 =

12. -3 • -19 =   13. 2 (-3) (-4) =   14. -2 • 3 • -6 =

15. -32 ÷ 8 =   16. -156 ÷ -3 =   17. $\frac{-136}{-8}$ =

18. $\frac{-32 \div -2}{16 \div 4}$ =   19. $\frac{5 \cdot -8}{-15 \div -3}$ =   20. $\frac{-40 \cdot -2}{-20 \div 2}$ =

1	
2	
3	
4	
5	
6	
7	
8	
9	
10	
11	
12	
13	
14	
15	
16	
17	
18	
19	
20	

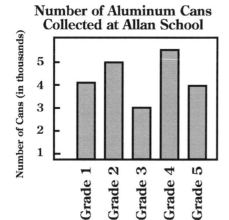

**Number of Aluminum Cans
Collected at Allan School**

1. Which grade collected the most cans?

2. How many more cans did grade 4 collect than grade 1?

3. How many cans did grades 4 and 5 collect altogether?

4. How many cans were collected in all?

5. Which grade collected the second-most number of cans?

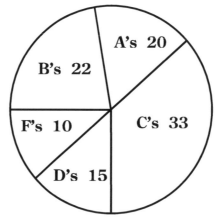

Fall Semester Science Grades

6. How many students got A's?

7. How many more C's than B's were there?

8. How many science students were there in all?

9. How many more C's and B's were there than A's and B's?

10. Which was the second largest group of grades?

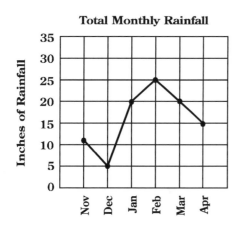

Total Monthly Rainfall

11. How many inches of rain fell in February?

12. How many more inches of rain fell in January than in November?

13. Which month's rainfall increased the most from the previous month?

14. What was the total amount of rain for February, March and April?

15. Which month's rainfall decreased the most from the previous month?

**Girl Scout Cookies
Sold During Fundraiser**

Troop 15	● ● ● ● ◖
Troop 5	● ● ●
Troop 16	● ● ● ◖
Troop 11	● ● ● ● ●
Troop 3	● ● ●

Each symbol represents 100 boxes

16. How many boxes of cookies did Troop 11 sell?

17. Which troop sold the second most cookies?

18. How many more boxes did Troop 11 sell than Troop 5?

19. If Troop 3 sells three times as many boxes of cookies next year than they did this year, how much will they sell next year?

20. How many boxes of cookies did Troop 5 and Troop 15 sell together?

1	
2	
3	
4	
5	
6	
7	
8	
9	
10	
11	
12	
13	
14	
15	
16	
17	
18	
19	
20	

132

1. A man drove his car 396 miles on Monday and 476 miles on Tuesday. How many miles did he drive altogether?

2. A plane traveled 2,175 miles in 5 hours. What was its average speed per hour?

3. A student took a test with 32 problems. If he got $\frac{5}{8}$ of the problems correct, how many problems did he get correct?

4. 8 pounds of nuts were put into bags which each held $1\frac{1}{3}$ pounds. How many bags were there?

5. Steve earned $15\frac{1}{2}$ dollars on Monday and $12\frac{3}{5}$ dollars on Tuesday. How much more did he make on Monday?

6. A train can travel 215 miles in one hour. At this pace, how far can it travel in .7 hours?

7. 8 pounds of apples cost $12.48. How much does one pound cost?

8. Hats were on sale for $23.50. If the regular price was $30.25, how much would you save buying it on sale?

9. Bill weighed 90 pounds, Steve weighed 120 pounds and Mary weighed 84 pounds. What was their average weight?

10. Bill earned $45.00 each day for 5 days and $65.50 the sixth day. How much did he earn in all?

11. A highway crew intended to pave 8 miles of road. If they paved $2\frac{1}{3}$ miles the first week and $1\frac{1}{2}$ miles the second week, how much as left?

12. Bill took a test with 24 questions. If he missed $\frac{1}{6}$ of them, how many did he get correct?

13. A car traveled 174 miles and averaged 29 miles per gallon of gas. If gas was $1.15 per gallon, how much did the trip cost?

14. To tune up his car, Stan bought six spark plugs for $1.15 each and a condenser for $4.29. How much did he spend altogether?

15. A living room in a house was 14 by 8 feet and the dining room was 10 by 12 feet. How many square feet of wall-to-wall carpet would be needed to cover both floors?

16. Cans of corn are 3 for $.76. How much would 12 cans cost?

17. A man can pay for a car in 36 payments of 150 dollars or pay 4,800 dollars cash. How much could he save by paying cash?

18. Jean baked a pie. If she gave $\frac{1}{2}$ to Steve and $\frac{1}{3}$ to Gina, how much of the pie did she have left?

19. For a party, Mr. Jones bought six pizzas for $9.95 each and 3 cases of soda for $7.15 per case. How much did he spend in all?

20. A man has a rectangular lot which is 36 feet by 20 feet. How many feet of fence are needed to enclose it? If each 4-foot section costs 55 dollars, how much will the fence cost?

1	
2	
3	
4	
5	
6	
7	
8	
9	
10	
11	
12	
13	
14	
15	
16	
17	
18	
19	
20	

Review Exercises

Note to students and teachers: This section will include necessary review problems from all types covered in this book. Here are some sample problems with which to get started.

1. $364 + 79 + 716 =$ 2. $705 - 269 =$ 3. $7 \times 326 =$

Helpful Hints

A set is a well-defined collection of objects. $A = \{1,2,3,4,5\}$ is read, "A is the set whose members are 1, 2, 3, 4, and 5. Each object in a set is called an element or member. **Infinite sets** are sets whose number of members is uncountable. **Example:** $A = \{1,2,3...\}$ **Finite sets** are sets whose number of members is countable. **Example:** $B = \{3,4,5\}$ **Disjoint sets** have no members in common. The **null set** or **empty set** is the set with no members, and is written as $\{ \}$ or $\emptyset$. **Equivalent sets** can be paired in a one-to-one correspondence. **Example:** $A = \{1,2,3,4\}$

$B = \{2,3,4,5\}$

$\cup$ = the universal set; the set that contains all the members.

$\in$ means "is a member of." $\notin$ means "is not a member of."

Use the information and examples given in the Helpful Hints to answer the following questions. Explain each answer in the space below.

S1. Is $A = \{2,4,6,8,10\}$ an infinite set?

S2. Is $B = \{2,4,6...\}$ a finite set?

1. Are $A = \{1,2,3\}$ and $B = \{3,4,5\}$ disjoint sets?

2. Are $C = \{0,1,2,3\}$ and $D = \{2,4,6,8\}$ equivalent sets?

3. List two disjoint sets.

4. List two equivalent sets.

For 5-10, list the members of each set.

5. {the odd numbers between 2 and 12}

6. {the even numbers less than 13}

7. {the whole numbers between 2 and 10}

8. {the multiples of five between 9 and 32}

9. The members common to $A = \{1,2,3,4,5\}$ and $B = \{1,3,5,7\}$

10. {the whole numbers greater than 7 and less than 13}

1.
2.
3.
4.
5.
6.
7.
8.
9.
10.
Score

Problem Solving

In a class of 38 students, one-half are girls. How many girls are there in the class?

Review Exercises

1. List two disjoint sets.   2. List two equivalent sets.   3. List an infinite set.

4. List a finite set.   5. Are A = {1,4,10} and B = {5,10,15} disjoint sets? Why?

6. Are C = {2,4,5} and D = {0,1,2,3} equivalent sets? Why?

Helpful Hints

Use the sets below for the following examples that pertain to subset, intersection, and union.

$$A = \{1,2,3\} \qquad B = \{0,1,2,3,4\} \qquad C = \{2,4,6,8,10\}$$

If **A** and **B** are sets and all the members of **A** are members of **B**, then **A** is a **subset** of **B** and is written $A \subset B$. **Example:** Is $A \subset B$? Yes, because all the members of **A** are members of **B**.

If **A** and **B** are sets then **A intersection B** is the set whose members are included in both sets **A** and **B**, and is written $A \cap B$. **Example:** Find $A \cap C$
$A \cap C = \{2\}$ (Two is the only member included in both **A** and **C**.)

If A and B are sets then **A union B** is the set whose members are included in **A** or **B**, or both **A** and **B**, and is written $A \cup B$. **Example:** Find $B \cup C$
$B \cup C = \{0,1,2,3,4,6,8,10\}$ ($B \cup C$ contains all members in **B**, **C**, or both **B** and **C**.)

Use the sets below to answer the questions on this page.
Explain in the space if necessary.

$$A = \{5,6,7\} \qquad B = \{1,2,3,4,5,6,7\} \qquad C = \{1,2,4,5,7,8\} \qquad D = \{1,2,4,6,8,10\}$$

S1. Is $A \subset B$? Why?   S2. Find $A \cap B$.

1. Find $B \cup C$.   2. Is $A \subset D$? Why?

3. List all subsets of A.   4. Find $B \cap C$.
   (Hint: there are eight of them.)

5. Find $C \cap D$.   6. Find $A \cup B$.

7. Find $B \cup D$.   8. Find $B \cap D$.

9. Are C and D equivalent sets? Why?

10. Are A and D disjoint sets? Why?

1.
2.
3.
4.
5.
6.
7.
8.
9.
10.

Problem Solving

Three weeks ago Jose sold seven of his baseball cards from his collection, and last week he bought 12 new cards. If he now has 85 cards, how many did he start with three weeks ago?

Score

Review Exercises

Use A = {1,2,3,5,6},   B = {2,4,8},   and C = {1,2,3,6}   to answer the following questions.

1. Find A ∩ B.   2. Find B ∪ C.   3. Find A ∩ C.   4. Find B ∩ C.

5. Are A and C equivalent sets? Why?   6. Is A an infinite set? Why?

Helpful Hints	**Integers are the set of whole numbers and their opposites.** 

Integers to the left of zero are negative and less than zero. Integers to the right of zero are positive and greater than zero. When two integers are on a number line, the one farthest to the right is greater. Hint: When adding integers, always find the sign of the answer first.

Examples: The sum of two negatives is a negative.

$-7 + -5 = -$ (the sign is negative) $\begin{array}{r} 7 \\ +5 \\ \hline \boxed{-12} \end{array}$

When adding a negative and a positive, the sign is the same as the integer farthest from zero. Then subtract.

$-7 + 9 = +$ (the sign is positive) $\begin{array}{r} 9 \\ -7 \\ \hline \boxed{+2} \end{array}$

S1. $-9 + 12 =$

S2. $-15 + -6 =$

1. $-15 + 29 =$

2. $-12 + -6 =$

3. $42 + -56 =$

4. $-15 + -16 =$

5. $8 + 32 =$

6. $-39 + 76 =$

7. $-96 + -72 =$

8. $73 + -86 =$

9. $-15 + -19 =$

10. $71 + -81 =$

1. _____
2. _____
3. _____
4. _____
5. _____
6. _____
7. _____
8. _____
9. _____
10. _____

Problem Solving	At 3:00 a.m. the temperature was -8°. By 6:00 a.m. the temperature was another 12° colder. What was the temperature at 6:00 a.m.?	Score

Review Exercises

1. $-16 + 9 =$

2. $-6 + 19 =$

3. $-26 + -13 =$

4. $-26 + 26 =$

5. Carefully define "set."

6. Carefully define "finite set."

| **Helpful Hints** | When adding more than two integers, group the negatives and positives separately, then add. | Examples:
$-6 + 4 + -5 =$
$-11 + 4 = -$
(sign is negative) | $\begin{array}{r} 11 \\ -4 \\ \hline 7 = \boxed{-7} \end{array}$ | $7 + -3 + -8 + 6 =$
$-11 + 13 = +$
(sign is positive) | $\begin{array}{r} 13 \\ -11 \\ \hline 2 = \boxed{+2} \end{array}$ |

S1. $-3 + 5 + -6 =$	S2. $-7 + 6 + -9 + 3 =$	1. $-3 + -4 + 5 =$
2. $7 + -6 + -8 =$	3. $-15 + 19 + -12 =$	4. $-6 + 9 + 7 + 4 =$
5. $-16 + 32 + -18 =$	6. $-13 + 16 + -8 + 15 =$	7. $-9 + -7 + -6 =$
8. $-3 + 7 + -8 + -9 =$	9. $-32 + 16 + -17 + 8 =$	10. $-76 + 25 + -33 =$

1.

2.

3.

4.

5.

6.

7.

8.

9.

10.

Score

Problem Solving

Alice started the week with no money. On Monday she earned $45.00. On Tuesday she spent $27.00. On Wednesday she earned $63.00. On Thursday she spent $26.00. How much money does she have left?

Review Exercises

Use the sets to answer problems 1 through 6.

A = {1,4,8,9,12},   B = {0,5,10,15}, and   C = {9,10,11,15} to answer the following questions.

1. Find A ∩ B.   2. Find A ∪ B.   3. Find B ∩ C.

4. Find A ∪ C.   5. Find A ∪ Ø.   6. Find B ∩ Ø.

Helpful Hints	To subtract an integer means to add to its opposite.	Examples: $-3 - -8 = $ 8 $-3 + 8 = +$ $- 3$ (sign is positive) $\overline{5}$ =(+5)	$8 - 10 = $ 10 $8 + -10 = -$ $- 8$ (sign is negative) $\overline{2}$ =(-2)	$6 - -7 = $ 7 $6 + 7 = +$ $+ 6$ (sign is positive) $\overline{13}$ =(+13)

S1. $-6 - 8 = $	S2. $6 - 9 = $	1. $3 - -9 = $	1.
			2.
			3.
2. $15 - 18 = $	3. $-16 - -25 = $	4. $-16 - 12 = $	4.
			5.
			6.
5. $32 - -14 = $	6. $-35 - 14 = $	7. $-6 - 4 = $	7.
			8.
			9.
8. $-64 - -53 = $	9. $-49 - 54 = $	10. $-63 - -78 = $	10.
			Score

Problem Solving	A boy jumped off a diving board that was 15 feet high. He touched the bottom of the pool that was 12 feet below the surface of the water. How far is it from the diving board to the bottom of the pool?

Review Exercises

1. -72 + 16 =

2. 55 + -33 =

3. -16 + -19 =

4. 7 - 16 =

5. -5 - 6 =

6. -5 - -9 =

Helpful Hints

The product of two integers with different signs is negative. The product of two integers with the same sign is positive. (• means multiply.)

Examples:

$7 \cdot -16 = -$ (sign is negative) $\begin{array}{r} 16 \\ \times\ 7 \\ \hline 112 \end{array}$ $= \boxed{-112}$

$-8 \cdot -7 = +$ (sign is positive) $\begin{array}{r} 8 \\ \times\ 7 \\ \hline 56 \end{array}$ $= \boxed{+56}$

When multiplying more than two integers, group them into pairs to simplify. An integer next to parenthesis means to multiply.

Examples:

$2 \cdot -3\ (-6) =$
$(2 \cdot -3)\ (-6) =$
$-6\ (-6) = +$
(sign is positive) $\begin{array}{r} 6 \\ \times\ 6 \\ \hline 36 \end{array}$ $= \boxed{+36}$

$-2 \cdot -3 \cdot 4 \cdot -2 =$
$(-2 \cdot -3)\ (4 \cdot -2) =$
$6 \cdot -8 = -$
(sign is negative) $\begin{array}{r} 6 \\ \times\ 8 \\ \hline 48 \end{array}$ $= \boxed{-48}$

S1. -3 × 16 =

S2. -18 • 7 =

1. -4 • -17 =

2. 16 × -4 =

3. -24 • -12 =

4. 23 × -16

5. -23 • 32 =

6. (-2) (-3) (-4) =

7. -8 (-1) • 1 (-4) =

8. 4 (-3) • 2 (-3) =

9. (-3) (-2) (3) (4) =

10. 10 (-11) (-3) =

1.
2.
3.
4.
5.
6.
7.
8.
9.
10.
Score

Problem Solving

An elevator started on the 28th floor. It went up seven floors, down 13 floors, and up nine floors. On what floor is the elevator located now?

Review Exercises

1. -27 + 16 =

2. -37 + -19 =

3. 7 - 9 =

4. -6 - -8 =

5. 5 • -7 =

6. -2 • -6 •3 =

Helpful Hints

The quotient of two integers with different signs is negative. The quotient of two integers with the same signs is positive. (HINT: Determine the sign, then divide.)
Examples:

$36 \div -4 = -$ (sign is negative) $4\overline{)36}$ $\dfrac{-36}{0} = $ (-9)

$\dfrac{-123}{-3} = +$ (sign is positive) $3\overline{)123}$ $\dfrac{-12}{3} = $ (+41)

Use what you have learned to solve problems like these.
Examples:

$\dfrac{-36 \div -9}{4 \div -2} = \dfrac{4}{-2} = $ (-2) (sign is negative)

$\dfrac{4 \times -8}{-8 \div 2} = \dfrac{-32}{-4} = $ (+8) (sign is positive)

S1. -36 ÷ 9 =

S2. $\dfrac{-90}{-15} =$

1. -64 ÷ 4 =

2. -336 ÷ -7 =

3. $\dfrac{-75}{-5} =$

4. 104 ÷ -4 =

5. $\dfrac{54 \div -9}{-18 \div -9} =$

6. $\dfrac{16 \div -2}{-1 \times -4} =$

7. $\dfrac{-75 \div -25}{-3 \div -1} =$

8. $\dfrac{42 \div -2}{-3 \bullet -7} =$

9. $\dfrac{45 \div -5}{-9 \div 3} =$

10. $\dfrac{-56 \div -7}{-36 \div -9} =$

1.
2.
3.
4.
5.
6.
7.
8.
9.
10.

Problem Solving

At midnight the temperature was 7°. By 2:00 a.m. the temperature had dropped 12°. By 4:00 a.m. it had dropped another 6°. What was the temperature at 4:00 a.m.?

Score

Reviewing All Integer Operations

1. $-9 + 7 =$	1.
2. $9 + -7 =$	2.
3. $-9 + -7 =$	3.
4. $-7 + -8 + 14 =$	4.
5. $-32 + 16 + 21 + -24 =$	5.
6. $7 - 9 =$	6.
7. $4 - -9 =$	7.
8. $-3 - 9 =$	8.
9. $-13 - 14 =$	9.
10. $16 - 17 =$	10.
11. $3 \cdot -16 =$	11.
12. $-4 \cdot -19 =$	12.
13. $2 \, (-7) \, (-4) =$	13.
14. $-2 \cdot 3 \, (-4) \cdot 2 =$	14.
15. $-36 \div 4 =$	15.
16. $-126 \div -3 =$	16.
17. $\dfrac{-128}{-8} =$	17.
18. $\dfrac{-36 \div 2}{24 \div -4} =$	18.
19. $\dfrac{6 \cdot -3}{-54 \div -6} =$	19.
20. $\dfrac{20 \cdot -3}{-30 \div -10} =$	20.

Review Exercises

Use the following sets to find the answers.

$$A = \{1,3,4,5,9\}, \quad B = \{1,2,4,6\}, \text{ and } \quad C = \{1,3,6,7\}$$

1. Find $A \cap B$.

2. Find $B \cup C$.

3. Find $B \cap C$.

4. Find $A \cup B$.

5. Find $A \cup C$.

6. Find $A \cup \varnothing$.

Helpful Hints

The rules for integers apply to positive and negative fractions.

Examples:

$-\dfrac{1}{2} + \dfrac{3}{5} =$

$-\dfrac{5}{10} + \dfrac{6}{10} = +$ (the sign is positive)

$\dfrac{6}{10} - \dfrac{5}{10} = \boxed{\dfrac{1}{10}}$

$-\dfrac{3}{5} + -\dfrac{1}{3} =$

$-\dfrac{9}{15} + -\dfrac{5}{15} = -$ (the sign is negative)

$\dfrac{9}{15} + \dfrac{5}{15} = \boxed{-\dfrac{14}{15}}$

$-\dfrac{3}{5} \times 1\dfrac{1}{2} = -\dfrac{4}{5}$

A negative multiplied or divided by a positive is negative.

$-\dfrac{3}{5} \times \dfrac{3}{2} = \boxed{-\dfrac{9}{10}}$

$-\dfrac{2}{3} \div -\dfrac{1}{2}$

A negative divided by a negative is a positive.

$-\dfrac{2}{3} \times -\dfrac{2}{1} = \dfrac{4}{3} = \boxed{1\dfrac{1}{3}}$

S1. $-\dfrac{1}{5} + \dfrac{1}{2} =$

S2. $\dfrac{1}{2} + -\dfrac{2}{5} =$

1. $\dfrac{1}{2} - \dfrac{3}{4} =$

2. $-\dfrac{2}{3} + -\dfrac{1}{2} =$

3. $-\dfrac{4}{5} \times 2\dfrac{1}{2} = -$

4. $\dfrac{5}{8} + -\dfrac{1}{4} =$

5. $-\dfrac{1}{3} - \dfrac{1}{4} =$

6. $-1\dfrac{2}{3} \times -1\dfrac{1}{2} =$

7. $\dfrac{3}{4} + \dfrac{1}{3} =$

8. $\dfrac{5}{8} + -\dfrac{1}{2} =$

9. $2\dfrac{1}{2} \div -\dfrac{1}{4} =$

10. $-\dfrac{1}{5} + \dfrac{2}{3} =$

1.

2.

3.

4.

5.

6.

7.

8.

9.

10.

Problem Solving

There are two sixth-grade classes. One has 35 students and another has 32 students. If a total of 17 sixth graders received A's, how many did not receive A's?

Score

Review Exercises

1. -75 + 16 =   2. -19 - 17 =   3. 16 × -4 =

4. -9 - 19 =   5. -36 ÷ -9 =   6. 2 • -7 × -2 =

Helpful Hints

The rules for integers apply to positive and negative decimals.

Example:

- .71 + .9 = + (the sign is positive)

$$\begin{array}{r} .90 \\ - \ .71 \\ \hline .19 \end{array}$$

Example:

-2.9 - 3.2 =

-2.9 + - 3.2 = - (the sign is negative)

$$\begin{array}{r} 2.9 \\ + \ 3.2 \\ \hline -6.1 \end{array}$$

Example:

-.5 × 1.23 = - (a negative multiplied or divided by a positive is a negative)

$$\begin{array}{r} 1.23 \\ \times \ \ .5 \\ \hline - .615 \end{array}$$

Example:

-3.12 ÷ - .3 = + (a negative multiplied or divided by a negative is a positive)

$$\begin{array}{r} 10.4 \\ .3\,\overline{)\,3.12} \end{array}$$

Work the following problems. If necessary, review the rules for integers.

S1. -3.21 + 2.3 =   S2. 5.15 ÷ -.5 =   1. -5.2 - 7.61 =

2. 5.63 + -2.46 =   3. -.7 × 6.12 =   4. 5.9 - -6.23

5. -7.11 ÷ -3 =   6. -.72 + .9 =   7. -2.13 × -.2 =

8. -6.2 + -.73 =   9. 5.2 + -3.19 =   10. -5.112 ÷ .3 =

1.
2.
3.
4.
5.
6.
7.
8.
9.
10.
Score

Problem Solving

Anna weighed 120.5 pounds. She lost 7.3 pounds and then gained back 4.8 pounds. How much does she weigh now?

Review Exercises

1. $-.3 + .7 =$

2. $-2.7 + -3.2 =$

3. $3 \times -2.6 =$

4. $-\frac{1}{2} + -\frac{1}{3} =$

5. $\frac{2}{5} + -\frac{1}{2} =$

6. $-\frac{1}{2} \times -1\frac{1}{5} =$

Helpful Hints

In the expression 5^3, the number 5 is called the **base** and the number 3 is called the **power** or **exponent**. The exponent tells how many times the base is to be multiplied by itself. In the example 5^3, you would multiply 5 three times: $5^3 = 5 \times 5 \times 5 = 125$. Negative numbers can have exponents: $(-2)^3 = (-2) \times (-2) \times (-2) = 4 \times (-2) = -8$. Any number to the power of 1 = the number. Any number to the power of 0 = 1.

Examples:

$3^4 = 3 \times 3 \times 3 \times 3$

$(-5)^4 = (-5) \times (-5) \times (-5) \times (-5)$

$5^1 = 5$

$= 9 \times 9$

$= 25 \times 25$

$6^0 = 1$

$= \boxed{81}$

$= \boxed{625}$

S1. $4^2 =$

S2. $-3^3 =$

1. $6^3 =$

2. $5^0 =$

3. $(-2)^4 =$

4. $2^5 =$

5. $7^1 =$

6. $8^3 =$

7. $(-1)^5 =$

8. $5^5 =$

9. $(-5)^3 =$

10. $(-3)^4 =$

1.	
2.	
3.	
4.	
5.	
6.	
7.	
8.	
9.	
10.	
Score	

Problem Solving

A certain number to the third power is equal to eight.
What is the number?

Review Exercises

1. $7^2 =$

2. $9^3 =$

3. $(-6)^2 =$

4. $5 + -6 + 8 + -3 =$

5. $7^0 =$

6. $9^1 =$

Helpful Hints	Many numbers can be written as exponents. **Examples:**

Many numbers can be written as exponents. **Examples:**

$5 \times 5 \times 5 \times 5 = 5^4$ $(-2) \times (-2) \times (-2) = (-2)^3$ $125 = 5^3$
$7 \times 7 \times 7 \times 7 \times 7 = 7^5$ $(-60) \times (-60) \times (-60) = (-60)^3$ $36 = 6^2$ or $(-6)^2$
$8 = 2^3$
$25 = 5^2$ or $(-5)^2$

Rewrite each of the following as an exponent.

S1. $12 \times 12 \times 12 =$ S2. $27 =$ 1. $2 \times 2 \times 2 \times 2 \times 2 \times 2 =$

2. $(-9) \times (-9) \times (-9) =$ 3. $16 \times 16 \times 16 \times 16 =$ 4. $49 =$

5. $100 =$ 6. $121 =$ 7. $(-1) \times (-1) \times (-1) \times (-1) =$

8. $32 =$ 9. $16 =$ 10. $9 \times 9 \times 9 \times 9 \times 9 \times 9 =$

1.
2.
3.
4.
5.
6.
7.
8.
9.
10.
Score

Problem Solving

A number to the third power is equal to -27. What is the number?

Review Exercises

1. $-36 \div 4 =$

2. $-9 - -6 =$

3. $-\dfrac{1}{3} + -\dfrac{1}{4} =$

4. $-2.7 + 6.3 =$

5. $-3.12 \div 3 =$

6. $\dfrac{3}{4} \times -\dfrac{1}{2} =$

Helpful Hints

$\sqrt{\ }$ is the symbol for **square root**.
$\sqrt{36}$ is read "the square root of 36."
The answer is the number that when multiplied by itself equals 36.
$\sqrt{36} = 6$, because $6 \times 6 = 36$.
$\sqrt{49} = 7$, because $7 \times 7 = 49$.
$\sqrt{81} = 9$, because $9 \times 9 = 81$.

Find the square roots of the following numbers.

S1. $\sqrt{25} =$ S2. $\sqrt{144} =$ 1. $\sqrt{16} =$

2. $\sqrt{121} =$ 3. $\sqrt{1} =$ 4. $\sqrt{900} =$

5. $\sqrt{100} =$ 6. $\sqrt{400} =$ 7. $\sqrt{169} =$

8. $\sqrt{9} =$ 9. $\sqrt{256} =$ 10. $\sqrt{1,600} =$

1.
2.
3.
4.
5.
6.
7.
8.
9.
10.

Score

Problem Solving

The product of -7 and 5 is added to -6.
Find the number.

Review Exercises

1. $6^2 =$

2. $(-2)^3 =$

3. write $6 \times 6 \times 6 \times 6$ as an exponent

4. $\sqrt{64} =$

5. $\sqrt{169} =$

6. $\sqrt{121} =$

Helpful Hints

Use what you have learned about exponents and square roots to solve the following problems.

Examples:

$$\frac{\sqrt{64}}{2^2} = \frac{8}{4} = 2 \qquad\qquad \frac{4^2}{2^3} = \frac{16}{8} = 2$$

$$\frac{3^3}{\sqrt{9}} = \frac{27}{3} = 9 \qquad\qquad \sqrt{36} \times 3^3 = 6 \times 27 = 162$$

Solve each of the following.

S1. $\sqrt{16} \times 3^2 =$

S2. $\dfrac{4^3}{\sqrt{64}} =$

1. $\dfrac{\sqrt{100}}{\sqrt{25}} =$

2. $\dfrac{\sqrt{64}}{(2^3)} =$

3. $\dfrac{5^3}{\sqrt{25}} =$

4. $2^3 \times \sqrt{121} =$

5. $3^2 \times 4^2 =$

6. $\dfrac{2^3 \times 4^2}{\sqrt{4}} =$

7. $\sqrt{81} \times \sqrt{36} =$

8. $\dfrac{2^4}{\sqrt{16}} =$

9. $2^2 \times 3^2 \times \sqrt{16} =$

10. $\dfrac{3^4}{\sqrt{81}} =$

1.	
2.	
3.	
4.	
5.	
6.	
7.	
8.	
9.	
10.	
Score	

Problem Solving

5 to the second power added to the square root of what number is equal to 34?

Reviewing Exponents and Square Roots

For 1-6, rewrite each as an exponent.

1. $13 \times 13 \times 13 \times 13 =$ 2. $2 \times 2 \times 2 \times 2 \times 2 \times 2 \times 2 =$

3. $64 =$ 4. $(-2) \times (-2) \times (-2) \times (-2) =$

5. $8 =$ 6. $100 =$

For 7-12, find the square root of each number.

7. $\sqrt{16} =$ 8. $\sqrt{64} =$ 9. $\sqrt{16 + 9} =$

10. $\sqrt{400}$ 11. $\sqrt{9}$ 12. $\sqrt{4} \times 9 =$

Solve each of the following

13. $\sqrt{36} + 4^2 =$ 14. $\dfrac{\sqrt{64}}{\sqrt{4}} =$ 15. $6^2 + 7^2 =$

16. $(4^2) \times (5^2) =$ 17. $\sqrt{49} \times \sqrt{81} =$ 18. $3^2 \times 5^2 \times \sqrt{9} =$

19. $\dfrac{5^3}{5} =$ 20. $\dfrac{\sqrt{100} \times \sqrt{25}}{\sqrt{25}} =$

1.
2.
3.
4.
5.
6.
7.
8.
9.
10.
11.
12.
13.
14.
15.
16.
17.
18.
19.
20.

Review Exercises

1. $7^2 =$

2. $\sqrt{36} =$

3. $-9 - -7 =$

4. $16 + -72$

5. $\dfrac{16 \div -2}{-4 \times -2} =$

6. $7^2 - 5^2 =$

Helpful Hints	It is necessary to follow the correct **order of operations** when simplifying an expression.
	  1. Evaluate within grouping symbols.
	  2. Eliminate all exponents.
	  3. Multiply and divide in order from left to right.
	  4. Add and subtract in order from left to right.

Examples:
$$3^2 (3 + 5) + 3$$
$$= 3^2 (8) + 3$$
$$= 9 (8) + 3$$
$$= 72 + 3$$
$$= 75$$

*A number next to a grouping symbol means multiply.

Sometimes there are no grouping symbols.
$$4 + 12 \times 3 - 8 \div 4$$
$$= 4 + 36 - 2$$
$$= 40 - 2$$
$$= 38$$

$$3 (2 + 1) = 3 \times (2 + 1)$$

Solve each of the following. Be sure to follow the correct order of operations.

S1. $5 + 9 \times 3 - 4 =$

S2. $8 + 3^2 \times 4 - 6 =$

1. $4 (6 + 2) - 5^2 =$

2. $(14 - 6) + 56 \div 2^3 =$

3. $5^2 + (15 + 3) \div 2 =$

4. $7 \times 4 - 9 \div 3 =$

5. $(3 \times 12) \div (9 \div 3) =$

6. $5^2 + 2^3 - 2 \times 3 =$

7. $12 - 6 \div 3 + 4 =$

8. $3^2 - 2^3 + 6 \div 2 =$

9. $(3 + 8 \div 2) \times (2 \times 6 \div 3) =$

10. $9 + [(4 + 5) \times 3] =$

1. _____

2. _____

3. _____

4. _____

5. _____

6. _____

7. _____

8. _____

9. _____

10. _____

Problem Solving	A running back gained 12 yards. The next play he lost 18 yards, and on the third play he gained five yards. What was his net gain or net loss?

Score _____

 Copyright © 2018 Richard W. Fisher

Review Exercises

Use the following sets to find the answers.

$$A = \{1,5,7,8,9\}, \quad B = \{2,4,6,8,10\}, \text{ and } \quad C = \{1,2,4,5\}$$

1. Find $B \cap C$.

2. Find $A \cap B$.

3. Find $C \cap \emptyset$.

4. Find $A \cup C$.

5. Find $B \cup C$.

6. Find $(A \cap B) \cup C$.

Helpful Hints	*Remember the correct order of operations: 1. Evaluate within grouping symbols. 2. Eliminate all exponents. 3. Multiply and divide in order from left to right. 4. Add and subtract in order from left to right.	**Examples:**

Examples:

$$14 \div 2 \times 3 + 4^2 - 1$$
$$= 14 \div 2 \times 3 + 16 - 1$$
$$= 7 \times 3 + 16 - 1$$
$$= 21 + 16 - 1$$
$$= 37 - 1$$
$$= 36$$

$$\frac{5(8-3) - 2^2}{3 + 2(3^2 - 7)} \rightarrow = \frac{25 - 4}{3 + 4}$$
$$= \frac{5(5) - 2^2}{3 + 2(9 - 7)} = \frac{21}{7}$$
$$= \frac{5(5) - 4}{3 + 2(2)} = 3$$

S1. $\{(2 + 4) \times 3 + 2\} \div 5 =$

S2. $\dfrac{7^2 - (-5 + 9)}{2(4^2 - 12) - 3} =$

1. $4^3 - 7(2 + 3) =$

2. $\dfrac{(12 - 3) + 3^2}{-7 + 2(4 + 1)} =$

3. $\dfrac{4^2 + 12}{5 + 3(2 + 1)} =$

4. $6(-3 + 9) + -4 =$

5. $\dfrac{10 + (2 + -6)}{4(2^3 - 6) + -2} =$

6. $63 \div 7 - 3 \times 2 + 4 =$

7. $3\{(2 + 7) \div 3 + 7\} \div 5 =$

8. $(12 + -3) + 75 \div 5^2 =$

9. $20 - 3^2 - 5 \times 2 + 6 =$

10. $3^2 + 2^3 + 14 \div 2 =$

1.	
2.	
3.	
4.	
5.	
6.	
7.	
8.	
9.	
10.	

Problem Solving

John started the week with $64. Each day, Monday through Friday, he spent $7 for lunch. How much money did he have left at the end of the week?

Score

Review Exercises

1. $3^3 =$

2. $5 + 3 \times 7 + 2 =$

3. $-7 - 6 =$

4. Carefully define "set."

5. $\frac{1}{2} \times -2\frac{1}{2} =$

6. $-.91 + .5 =$

Helpful Hints	For any real numbers a, b, and c, the following properties are true:		**Examples:**
	1. Identity Property of Addition	$0 + a = a$	$0 + 2 = 2$
	2. Identity Property of Multiplication	$1 \times a = a$	$1 \times 7 = 7$
	3. Inverse Property of Addition	$a + (-a) = 0$	$5 + (-5) = 0$
	4. Inverse Property of Multiplication	$a \times \frac{1}{a} = 1 \quad (a \neq 0)$	$6 \times \frac{1}{6} = 1$
	5. Associative Property of Addition	$(a + b) + c = a + (b + c)$	$(2 + 3) + 4 = 2 + (3 + 4)$
	6. Associative Property of Multiplication	$(a \times b) \times c = a \times (b \times c)$	$(2 \times 3) \times 4 = 2 \times (3 \times 4)$
	7. Commutative Property of Addition	$a + b = b + a$	$5 + 6 = 6 + 5$
	8. Commutative Property of Multiplication	$a \times b = b \times a$	$4 \times 3 = 3 \times 4$
	9. Distributive Property	$a \times (b + c) = a \times b + a \times c$	$5 \times (3 + 2) = 5 \times 3 + 5 \times 2$

Name the property that is illustrated.

S1. $7 + 9 = 9 + 7$ **S2.** $3 \times (7 + 4) = 3 \times 7 + 3 \times 4$ **1.** $7 + (-7) = 0$

2. $3 \times (4 \times 5) = (3 \times 4) \times 5$ **3.** $0 + (-6) = -6$ **4.** $5 \times \frac{1}{5} = 1$

5. $9 + (6 + 5) = 9 + (5 + 6)$ **6.** $9 \times 7 = 7 \times 9$ **7.** $(6 + 5) + 7 = 6 + (5 + 7)$

8. $1 \times 7 = 7$ **9.** $3 \times 2 + 3 \times 4 = 3 \times (2 + 4)$ **10.** $16 + (-16) = 0$

1.

2.

3.

4.

5.

6.

7.

8.

9.

10.

Problem Solving

Five times a certain number is equal to 95. Find the number.

Score

Review Exercises

1. $3 + 6 \times 2 - 4 =$

2. $3 (5 + 2) - 2^2 =$

3. $3 \times 4 - 6 \div 3 =$

4. $2^2 + 3^3 - 2 \times 4 =$

5. $(4 + 4 \div 2) \times (2 \times 10 \div 2) =$

6. $3 [(4 + 3) \times 2] =$

| **Helpful Hints** | Use what you have learned to solve the following problems.
 Example: Use the indicated property to complete the statement with the correct answer.
 Inverse Property of Addition: $27 + (\;) = 0$ answer: -27
 Distributive Property: $4 (5 + 7) =$ answer: $4 \times 5 + 4 \times 7$ |

Use the indicated property to complete the statement with the correct answer.

S1. Associative Property of Addition: $(3 + 7) + 9 =$

S2. Commutative Property of Multiplication: $7 \times 15 =$

1. Inverse Property of Multiplication: $9 \times (\;) = 1$

2. Distributive Property: $3 \times (6 + 2) =$

3. Commutative Property of Addition: $9 + 12 =$

4. Associative Property of Multiplication: $3 \times (9 \times 5) =$

5. Distributive Property: $3 \times 5 + 3 \times 7 =$

6. Inverse Property of Addition: $9 + (\;) = 0$

7. Identity Property of Multiplication: $7 \times (\;) = 7$

8. Inverse Property of Multiplication: $\frac{1}{5} (\;) = 1$

9. Associative Property of Addition: $3 + (5 + 6) =$

10. Distributive Property: $6 \times (4 + (-2)) =$

1.
2.
3.
4.
5.
6.
7.
8.
9.
10.

Score

Problem Solving   Mr. Andrews rents a car for one day. He pays $30 per day for the rental plus $.30 per mile he drives. How much will the total price of the rental car be if he drives 40 miles?

Review Exercises

1. $\sqrt{100}$

2. $5^3 =$

3. $\dfrac{\sqrt{36} \times \sqrt{49}}{2} =$

4. $-3 \times -5 \times -6 =$

5. $-2\dfrac{1}{2} \div -\dfrac{1}{2} =$

6. $-\dfrac{1}{3} + -\dfrac{1}{2} =$

Helpful Hints

Scientific notation is used to express very large and very small numbers. A number in scientific notation is expressed as the product of two factors. The first factor is a number between 1 and 10 and the second factor is a power of 10 as in the examples **2.346×10^5** and **3.976×10^{-7}**.

Example for a large number: Change 157,000,000,000 to scientific notation.
Move the decimal between the 1 and the 5. Since the decimal has moved 11 places to the **left**, the answer is 1.57×10^{11}.

Example for a small number: Change .0000000468 to scientific notation.
Move the decimal between the 4 and the 6. Since the decimal has moved eight places to the **right**, the answer is 4.68×10^{-8}.

Change the following numbers to scientific notation.

S1. 2,360,000,000

S2. .000000149

1. 653,000,000,000

2. 159,700

3. 106,000,000

4. .000007216

5. 1,096,000,000

6. .001963

7. .00000000016

8. .0000000008

9. 7,000,000,000,000

10. .0000001287

1.	
2.	
3.	
4.	
5.	
6.	
7.	
8.	
9.	
10.	
Score	

Problem Solving

Light travels at 186,000 miles per second. Write this speed in scientific notation.

Copyright © 2018 Richard W. Fisher

Review Exercises

1. Change 123,000 to scientific notation.

2. Change .000321 to scientific notation.

3. Which property of numbers is illustrated?
$3 \times 5 + 3 \times 7 = 3 \times (5 + 7)$

4. $-9 - 8 =$

5. $\dfrac{36 \div -3}{-16 \div -4} =$

6. $2^3 + 3^3 =$

Helpful Hints

It is easy to change numbers in scientific notation to conventional numbers.

Examples:

Change 3.458×10^8 to a conventional number.
Move the decimal eight places to the right. The answer is 345,800,000.

Change 4.5677×10^{-7} to a conventional number.
Move the decimal seven spaces to the left. The answer is .00000045677.

Change each number in scientific notation to a conventional number.

S1. 7.032×10^6

S2. 5.6×10^{-5}

1. 2.3×10^5

2. 9.13×10^{-8}

3. 1.2362×10^{-5}

4. 5.17×10^{11}

5. 1.127×10^3

6. 3.012×10^{-3}

7. 6.67×10^6

8. 2.1×10^4

9. 7×10^{-8}

10. 8×10^6

1.	
2.	
3.	
4.	
5.	
6.	
7.	
8.	
9.	
10.	

Problem Solving

The distance to the sun is approximately 9.3×10^7 miles. Change this distance to a conventional number.

Score

Review Exercises

1. Change 123,000 to scientific notation.

2. Change .0000056 to scientific notation.

3. Change 2.76×10^6 to a conventional number.

4. Change 3.75×10^{-5} to a conventional number.

5. List two equivalent sets.

6. List two disjoint sets.

Helpful Hints

A **ratio** compares two numbers or groups of objects.

Example: ○ ○ ○   For every three circles there are four squares.
□ □ □ □

The ratio can be written in the following ways:
3 to 4, 3 : 4, and $\frac{3}{4}$. Each of these is read as "three to four."

*Ratios are often written in fraction form. The first number mentioned is the numerator. Ratios that are expressed as fractions can be reduced to lowest terms.

Write each of the following ratios as a fraction reduced to lowest terms.

S1. 5 nickels to 3 dimes

S2. 18 horses to 4 cows

1. 7 to 2

2. 6 children to 5 adults

3. 30 books to 25 pencils

4. 15 bats to 3 balls

5. 24 to 20

6. 16 to 12

7. 7 dimes to 3 pennies

8. 6 chairs to 4 desks

9. 4 cats to 8 dogs

10. 9 : 3

1.
2.
3.
4.
5.
6.
7.
8.
9.
10.

Problem Solving

A team won 24 games and lost 10. Write the ratio of games won to games lost as a fraction reduced to lowest terms.

Score

Review Exercises

1. Write .00027 in scientific notation.

2. Write 2,916,000 in scientific notation.

3. Write 7.21×10^5 as a conventional number.

4. Write 6.23×10^{-5} as a conventional number.

5. $(-3) \times 2 \times (-5) =$

6. $-.264 \div .2 =$

Helpful Hints

Two equal ratios can be written as a **proportion**.

Example: $\frac{4}{6} = \frac{2}{3}$ In a proportion, the cross products are equal.

Examples: Is $\frac{3}{4} = \frac{5}{6}$ a proportion? To find out, cross multiply.

$3 \times 6 = 18$, $4 \times 5 = 20$, $18 \neq 20$. It is not a proportion.

Is $\frac{6}{9} = \frac{8}{12}$ a proportion? To find out, cross multiply.

$6 \times 12 = 72$, $9 \times 8 = 72$, $72 = 72$. It is a proportion.

Cross multiply to determine whether each of the following is a proportion.

S1. $\frac{2}{5} = \frac{6}{15}$

S2. $\frac{18}{24} = \frac{4}{5}$

1. $\frac{7}{14} = \frac{3}{6}$

2. $\frac{5}{3} = \frac{14}{9}$

3. $\frac{18}{2} = \frac{27}{3}$

4. $\frac{4}{5} = \frac{12}{15}$

5. $\frac{15}{20} = \frac{6}{8}$

6. $\frac{5}{2} = \frac{11}{4}$

7. $\frac{2}{5} = \frac{12}{30}$

8. $\frac{3}{1.3} = \frac{9}{3.5}$

9. $\frac{1/4}{4} = \frac{1/2}{8}$

10. $\frac{5}{8} = \frac{6}{7}$

1.	
2.	
3.	
4.	
5.	
6.	
7.	
8.	
9.	
10.	
Score	

Problem Solving

A whole number to the power of three, added to five, equals 13. Find the whole number.

Review Exercises

1. Write 25 to 15 as a fraction in lowest terms.

2. Is $\dfrac{4}{5} = \dfrac{8}{10}$ a proportion? Why?

3. Is $\dfrac{2}{5} = \dfrac{5}{7}$ a proportion? Why?

4. $\sqrt{49} + 3^2 =$

5. $4^3 - 2^4 =$

6. $-225 + 500 =$

Helpful Hints

It is easy to find the missing number in a proportion.

Examples: Solve each proportion.

$\dfrac{4}{n} = \dfrac{2}{3}$

First, cross multiply: $\quad 2 \times n = 4 \times 3$
$\qquad\qquad\qquad\qquad\quad 2 \times n = 12$

Next, divide 12 by 2: $\quad 2\overline{)12} \enspace \dfrac{6}{}$ $\enspace \boxed{n = 6}$

$\dfrac{4}{5} = \dfrac{y}{7}$

First, cross multiply: $\quad 5 \times y = 4 \times 7$
$\qquad\qquad\qquad\qquad\quad 5 \times n = 28$

Next, divide 28 by 5: $\quad 5\overline{)28} \enspace 5\tfrac{3}{5}$ $\enspace \boxed{y = 5\tfrac{3}{5}}$

Find the missing number in each proportion.

S1. $\dfrac{3}{15} = \dfrac{n}{5}$

S2. $\dfrac{4}{7} = \dfrac{x}{28}$

1. $\dfrac{n}{4} = \dfrac{12}{16}$

2. $\dfrac{x}{40} = \dfrac{5}{100}$

3. $\dfrac{1}{3} = \dfrac{14}{y}$

4. $\dfrac{n}{4} = \dfrac{8}{5}$

5. $\dfrac{15}{20} = \dfrac{n}{8}$

6. $\dfrac{7}{n} = \dfrac{3}{9}$

7. $\dfrac{27}{3} = \dfrac{n}{2}$

8. $\dfrac{n}{2} = \dfrac{7}{5}$

9. $\dfrac{n}{1.4} = \dfrac{6}{7}$

10. $\dfrac{7}{n} = \dfrac{21}{7}$

1.	
2.	
3.	
4.	
5.	
6.	
7.	
8.	
9.	
10.	
Score	

Problem Solving

The temperature at midnight is -12°. By 6:00 a.m., the temperature has dropped another 20°. What is the temperature at 6:00 a.m.?

Review Exercises

1. Is $\frac{3}{4} = \frac{9}{12}$ a proportion? Why?

2. Solve the proportion:
$$\frac{n}{12} = \frac{5}{2}$$

3. Solve the proportion:
$$\frac{5}{6} = \frac{10}{n}$$

4. Write 234,000,000 in scientific notation.

5. Write .00235 in scientific notation.

6. Write 7.2×10^5 as as a conventional number.

Helpful Hints

Ratios and **proportions** can be used to solve problems.

Example: A car can travel 384 miles in six hours. How far can the car travel in eight hours?

First set up a proportion. $\frac{384 \text{ miles}}{6 \text{ hours}} = \frac{n \text{ miles}}{8 \text{ hours}}$ → Next, divide by six.

Next, cross multiply. $6 \times n = 8 \times 384$
$6 \times n = 3,072$

$6\,\overline{)3072}$ → 512 $n = 512$

The car can travel 512 miles in eight hours.

Use a proportion to solve each problem.

S1. A car can travel 85 miles on five gallons of gas. How far can the car travel on 12 gallons of gas?

S2. If two pounds of beef cost $4.80, how much will five pounds cost?

1. A car can travel 100 miles on five gallons of gas. How many gallons will be needed to travel 40 miles?

2. Two pounds of chicken cost $7. How much will five pounds cost?

3. In a class, the ratio of boys to girls is four to three. If there are 20 boys in the class, how many girls are there?

4. A runner takes three hours to go 24 miles. At this rate, how far could he run in five hours?

5. Seven pounds of nuts cost $5. How many pounds of nuts can you buy with $2?

1.	
2.	
3.	
4.	
5.	
Score	

Problem Solving

At 6:00 a.m. the temperature was -16°. By noon the temperature had risen 28°. What was the temperature at noon?

Reviewing Ratios and Proportions

For 1-3, write each ratio as a fraction reduced to lowest terms.

1. 12 to 4

2. 24 to 10

3. 16 to 6

For 4-6, determine whether each is a proportion and why.

4. $\dfrac{12}{15} = \dfrac{24}{30}$

5. $\dfrac{7}{8} = \dfrac{8}{9}$

6. $\dfrac{5}{3} = \dfrac{15}{9}$

For 7-15, solve each proportion.

7. $\dfrac{12}{15} = \dfrac{n}{5}$

8. $\dfrac{1}{3} = \dfrac{11}{n}$

9. $\dfrac{1}{20} = \dfrac{n}{100}$

10. $\dfrac{5}{7} = \dfrac{25}{n}$

11. $\dfrac{3}{4} = \dfrac{n}{6}$

12. $\dfrac{15}{20} = \dfrac{n}{12}$

13. $\dfrac{10}{100} = \dfrac{2}{n}$

14. $\dfrac{x}{5} = \dfrac{9}{15}$

15. $\dfrac{3}{16} = \dfrac{n}{48}$

For 16-20, use a proportion to solve each problem.

16. If four pounds of pork cost $4.80, how much will seven pounds cost?

17. In a class the ratio of girls to boys is two to three. If there are 20 girls, how many boys are in the class?

18. A cyclist can travel 42 miles in three hours. How far can he travel in five hours?

19. A car can travel 120 miles on five gallons of gas. How many gallons will be needed to travel 48 miles?

20. If six pounds of nuts cost $18, how many pounds of nuts can you buy with $12?

1.
2.
3.
4.
5.
6.
7.
8.
9.
10.
11.
12.
13.
14.
15.
16.
17.
18.
19.
20.

Review Exercises

1. Solve the proportion:
$$\frac{7}{6} = \frac{n}{18}$$

2. Solve the proportion:
$$\frac{n}{3} = \frac{6}{5}$$

3. $3 \times 2 + 6 \div 2 =$

4. $4^2 + (5 \times 2) \div 5 =$

5. $4^2 + 2^2 + 12 \div 2 =$

6. $5(-2 + -6) + 7 =$

Helpful Hints

Percent means "**per hundred**" or "**hundredths.**"

Percents can be expressed as decimals and as fractions.
The fraction form may sometimes be reduced to its lowest terms.

Examples: $25\% = .25 = \frac{25}{100} = \frac{1}{4}$ $8\% = .08 = \frac{8}{100} = \frac{2}{25}$

Change each percent to a decimal and to a fraction reduced to its lowest terms.

S1. $20\% =$ _____

S2. $9\% =$ _____

1. $16\% =$ _____

2. $6\% =$ _____

3. $75\% =$ _____

4. $40\% =$ _____

5. $1\% =$ _____

6. $45\% =$ _____

7. $12\% =$ _____

8. $5\% =$ _____

9. $50\% =$ _____

10. $13\% =$ _____

1.
2.
3.
4.
5.
6.
7.
8.
9.
10.

Problem Solving

95% of the students enrolled in a school are present.
What fraction are present? (Reduce to lowest terms.)

Score

Review Exercises

1. Change 80% to a decimal.

2. Change 7% to a decimal.

3. Change 25% to a fraction reduced to lowest terms.

4. 156
   × .7

5. 400
   × .32

6. 300
   × .06

Helpful Hints

To find the **percent of a number**, you may use either fractions or decimals. Use what is the most convenient.

Example: Find 25% of 60.

$.25 \times 60$

$$\begin{array}{r} 60 \\ \times\ .25 \\ \hline 300 \\ 120 \\ \hline 15.00 \end{array}$$

OR

$$\frac{25}{100} = \frac{1}{4}$$

$$\frac{1}{4} \times \frac{60^{15}}{1} = \frac{15}{1} = \boxed{15}$$

1.
2.
3.
4.
5.
6.
7.
8.
9.
10.

S1. Find 70% of 25.

S2. Find 50% of 300.

1. Find 6% of 72.

2. Find 60% of 85.

3. Find 25% of 60.

4. Find 45% of 250.

5. Find 10% of 320.

6. Find 40% of 200.

7. Find 4% of 250.

8. Find 90% of 240.

9. Find 75% of 150.

10. Find 2% of 660.

Problem Solving

Arlene took a test with 40 questions. If she got a score of 85% correct, how many problems did she get correct?

Score

Review Exercises

1. Find 15% of 310.

2. Find 20% of 120.

3. $8\overline{\smash{)}6}$

4. Change .7 to a percent.

5. Find .9 of 150.

6. $.05\overline{\smash{)}30}$

Helpful Hints

When finding the **percent**, first write a fraction, change the fraction to a decimal, and then change the decimal to a percent.

Examples: 4 is what percent of 16?

$\dfrac{4}{16} = \dfrac{1}{4}$

$\begin{array}{r} .25 = \boxed{25\%} \\ 4\,\overline{)1.00} \\ -\ 8\!\downarrow \\ \overline{20} \\ -\ 20 \\ \hline 0 \end{array}$

5 is what percent of 25?

$\dfrac{5}{25} = \dfrac{1}{5}$

$\begin{array}{r} 20 = \boxed{20\%} \\ 5\,\overline{)1.00} \\ -\ 1\ 0 \\ \hline 00 \end{array}$

S1. 3 is what percent of 12?

S2. 15 is what percent of 20?

1. 7 is what percent of 28?

2. 20 is what percent of 25?

3. 40 = what percent of 80?

4. 18 is what percent of 20?

5. 12 is what percent of 20?

6. 9 is what percent of 12?

7. 15 = what percent of 20?

8. 24 is what percent of 32?

9. 400 is what percent of 500?

10. 19 is what percent of 20?

1.
2.
3.
4.
5.
6.
7.
8.
9.
10.

Problem Solving

A rancher had 800 cows. He sold 600 of them. What percent of the cows did he sell?

Score

Review Exercises

1. Find 4% of 80.

2. Find 40% of 80.

3. Twelve is what percent of 16?

4. 45 is what percent of 50?

5. 52 - 1.96 =

6. $.06\overline{\smash{)}12}$

Helpful Hints

To find the **whole** when the **part** and the **percent** are known, simply change the equal sign (" = ") to the division sign (" ÷ ").

Examples: 6 = 25% of what number?   Twelve is 80% of what?
6 ÷ 25% (Change = to ÷.)   12 ÷ 80% (Change = to ÷.)
6 ÷ .25 (Change % to decimal.)   12 ÷ .8 (Change % to decimal.)

⟨24.⟩   ⟨15.⟩
$.25\overline{\smash{)}6.00}$ * Be careful to move $.8\overline{\smash{)}12.0}$
                                  decimal points properly.

S1. 5 = 25% of what?	S2. Six is 20 % of what?	1.
		2.
1. 12 = 25% of what?	2. 32 = 40 % of what?	3.
3. Five is 20% of what?	4. 3 = 75% of what?	4.
		5.
5. Twelve is 80% of what?	6. 8 = 40% of what?	6.
7. 15 is 25% of what?	8. Fifteen is 20% of what?	7.
		8.
9. 9 is 20% of what?	10. 25 is 20% of what?	9.
		10.

Problem Solving

There are 15 girls in a class. If this is 60% of the class, how many students are there in the class?

Score

Review Exercises

1. Change $\frac{72}{100,000}$ to a decimal.   2. Change 2.0019 to a mixed numeral.   3. Change $\frac{9}{15}$ to a percent.

4. $\frac{3}{5} \times 25 =$ 5. $8\overline{)\,.168}$ 6. $.03\overline{)\,2.4}$

Helpful Hints

Use what you have learned to solve the following word problems. **Examples:**

A man earns \$300 and spends 40% of it. How much does he spend?

Find 40% of 300.

$$\begin{array}{r} 300 \\ \times\,.4 \\ \hline 120.0 \end{array}$$

He spends \$120.

In a class of 25 students, 15 are girls. What % are girls?

15 = what % of 25

$$\frac{15}{25} = \frac{3}{5}$$

$$5\overline{)3.00}\;\;.60$$

60% are girls.

Five students got A's on a test. This is 20% of the class. How many are in the class?

5 = 20% of what?
5 ÷ .2

$$.2\overline{)5.0}\;\;25.$$

25 are in the class.

S1. On a test with 25 questions, Al got 80% correct. How many questions did he get correct?

S2. A player took 12 shots and made 9. What percent did he make?

1. A girl spent \$5. This was 20% of her earnings. How much were her earnings?

2. Buying a \$8,000 car requires a 20% down payment. How much is the down payment?

3. 3 = 10% of what?

4. A team played 20 games and won 18. What % did they win?

5. A farmer sold 50 cows. If this was 20% of his herd, how many cows were in his herd?

6. 20 = 80% of what?

7. Paul wants a bike that costs \$400. If he has saved 60% of this amount, how much has he saved?

8. There are 400 students in a school. Fifty-five percent are girls. How many boys are there?

9. 12 is what % of 60?

10. Kelly earned 300 dollars and put 70% of it into the bank. How much did she put into the bank?

1. _____
2. _____
3. _____
4. _____
5. _____
6. _____
7. _____
8. _____
9. _____
10. _____

Problem Solving

Nacho's monthly income is \$4,800. What is his annual income? (Hint: How many months are there in a year?)

Score

Review Exercises

1. $7.68 + 19.7 + 5.364 =$

2. $\begin{array}{r} 7.123 \\ - 4.765 \\ \hline \end{array}$

3. $\begin{array}{r} 3.14 \\ \times\ 7 \\ \hline \end{array}$

4. $\begin{array}{r} .208 \\ \times\ .06 \\ \hline \end{array}$

5. $3\overline{)1.44}$

6. $.15\overline{)1.215}$

Helpful Hints

Use what you have learned to solve the following problems.

*Refer to the examples on the previous page if necessary.

S1. Find 20% of 150.	S2. 6 is 20% of what?	1.
1. 8 is what % of 40?	2. Change $\frac{18}{20}$ to a percent.	2.
3. A school has 600 students. If 5% are absent, how many students are absent?	4. A quarterback threw 24 passes and 75% of them were caught. How many were caught?	3.
		4.
5. Riley has 250 marbles in his collection. If 50 of them are red, what percent of them are red?	6. A team played 60 games and won 45 of them. What % did they win?	5.
		6.
7. There are 50 sixth graders in a school. This is 20% of the school. How many students are in the school?	8. A coat is on sale for $20. This is 80% of the regular price. What is the regular price?	7.
		8.
9. Steve has finished $\frac{3}{5}$ of his test. What percent of the test has he finished?	10. Alex wants to buy a computer priced at $640. If sales tax is 8%, what is the total cost of the computer?	9.
		10.
		Score

Problem Solving

Ann took five tests and scored a total of 485 points. What was her average score?

Reviewing Percents

Change numbers 1 - 5 to percents.

1. $\dfrac{13}{100} =$ 2. $\dfrac{3}{100} =$ 3. $\dfrac{7}{10} =$ 4. $.19 =$ 5. $.6 =$

Change numbers 6 - 8 to a decimal and a fraction expressed in lowest terms.

6. $8\% = .\underline{\quad} = \underline{\quad\quad}$ 7. $18\% = .\underline{\quad} = \underline{\quad\quad}$ 8. $80\% = .\underline{\quad} = \underline{\quad\quad}$

Solve the following problems. Label the word problem answers.

9. Find 3% of 74.

10. Find 40% of 320.

11. 20 is what percent of 25?

12. 15 is what percent of 20?

13. 3 = 25% of what?

14. 15 = 20% of what?

15. Change $\dfrac{16}{20}$ to a %.

16. Change $\dfrac{3}{5}$ to a percent.

17. 640 students attend Lincoln School. If 40% of the students are girls, how many girls attend Lincoln School?

18. A team played 40 games. If they won 65% of them, how many games did the team win?

19. A pitcher threw 40 pitches. If 30 were strikes, what percent were strikes?

20. Thirty students attended an assembly. This was 20% of the seventh grade. How many students are there in the seventh grade?

1.
2.
3.
4.
5.
6.
7.
8.
9.
10.
11.
12.
13.
14.
15.
16.
17.
18.
19.
20.

Review Exercises

1. Change .7 to a percent.

2. Change $\frac{4}{5}$ to a percent.

3. Change .12 to a fraction reduced to lowest terms.

4. Find 6% of 200.

5. Three is what percent of 12?

6. 5 = 20% of what?

Helpful Hints

A **factor** of a whole number is a whole number that divides into it evenly, without a remainder.

Examples: Find all factors of 20.

$1 \times 20 = 20$ $2 \times 10 = 20$ $4 \times 5 = 20$
All the factors of 20 are: 1, 20, 2, 20, 4, 5

Find all factors of 84.

$1 \times 84 = 84$ $2 \times 42 = 84$ $3 \times 28 = 84$
$4 \times 21 = 84$ $6 \times 14 = 84$ $7 \times 12 = 84$

All the factors of 84 are: 1, 84, 2, 42, 3, 28, 4, 21, 6, 14, 7, 12

Find all the factors of each number.

S1. 30

S2. 36

1. 100

2. 42

3. 70

4. 81

5. 50

6. 40

7. 75

8. 90

9. 20

10. 28

1.	
2.	
3.	
4.	
5.	
6.	
7.	
8.	
9.	
10.	
Score	

Problem Solving

A test contained 60 questions. If a student's score was 90%, how many questions did he get correct?

Review Exercises

1. -9 - 6 + -3 =

2. -3 × -2 • 4 =

3. $\sqrt{121} + \sqrt{81}$

4. Solve the proportion.
$$\frac{3}{4} = \frac{n}{10}$$

5. 3 = 20% of what?

6. Two is what % of eight?

Helpful Hints

The **greatest common factor** is the largest factor that two or more numbers have in common.

Example: Find the greatest common factor of 12 and 16.

Find the factors of each number: 12: 1, 2, 3,④,6, 12
16: 1, 2,④,8, 16   greatest common factor = ④

* "Greatest common factor" is abbreviated as GCF.

Find the greatest common factor of each pair of numbers.

S1. 8 and 10

S2. 12 and 20

1. 6 and 8

2. 12 and 15

3. 42 and 56

4. 64 and 80

5. 100 and 120

6. 90 and 70

7. 45 and 25

8. 60 and 72

9. 48 and 36

10. 20 and 40

1.

2.

3.

4.

5.

6.

7.

8.

9.

10.

Problem Solving

Light travels approximately 5.879×10^{12} miles in one year. Write the distance travelled as a conventional number.

Score

Review Exercises

1. Write .0000012 in scientific notation.

2. Write 496,000,000 in scientific notation.

3. Write 1.32×10^7 as a conventional number.

4. Write 4.64×10^{-6} as a conventional number.

5. Find all the factors of 60.

6. Find the GCF (greatest common factor) of 36 and 40.

Helpful Hints

A **multiple** of a number is the product of that number and any whole number.

The multiples of a number can be found by multiplying it by 0, 1, 2, 3, 4, and so on.

Example: Find the first six multiples of 3.

3: 0, 3, 6, 9, 12, 15

These are found by multiplying 3 by 0, 1, 2, 3, 4, and 5.

Complete the list of multiples for each number.

S1. 2: 0, 2, ☐, ☐, ☐, ☐

S2. 6: ☐, 6, ☐, ☐, 24, ☐

1. 5: 0, 5, ☐, ☐, ☐, ☐

2. 3: ☐, 3, ☐, 9, ☐, ☐

3. 10: ☐, 10, 20, ☐, ☐, ☐

4. 4: ☐, ☐, ☐, 12, 16, 20

5. 11: 0, 11, ☐, 33, ☐, 55

6. 8: 0, 8, 16, ☐, ☐, ☐

7. 20: 0, 20, 40, ☐, ☐, ☐

8. 7: 0, 7, ☐, 21, ☐, ☐

9. 30: 0, 30, 60, ☐, ☐, ☐

10. 9: 0, 9, 18, ☐, 36, ☐

1.
2.
3.
4.
5.
6.
7.
8.
9.
10.
Score

Problem Solving

A pitcher threw 30 pitches that were strikes.
This was 25% of all the pitches thrown.
How many pitches were thrown by the pitcher?

Review Exercises

1. List all the factors of 30.

2. Find the GCF of 32 and 60.

3. List the first six multiples of eight.

4. Find 6% of 50.

5. 3 is what % of 12?

6. 7 = 20% of what?

Helpful Hints

The **least common multiple** of two or more whole numbers is the smallest whole number, other than zero, that they all divide into evenly.

Examples: The least common multiple of:

2 and 3 is 6   4 and 6 is 12   3 and 9 is 9

* Least common multiple is abbreviated as LCM.

Find the least common multiple of each pair of numbers.

S1. 3 and 4	S2. 6 and 8

1. 3 and 5

2. 6 and 10

3. 12 and 20

4. 10 and 15

5. 12 and 18

6. 15 and 60

7. 16 and 12

8. 8 and 20

9. 9 and 12

10. 12 and 30

1.
2.
3.
4.
5.
6.
7.
8.
9.
10.
Score

Problem Solving

A CD costs $12. If the sales tax is 8%, what is the total cost of the CD?

Reviewing Number Theory

For 1-6, find all factors for each number.

1. 24   2. 16   3. 32

4. 28   5. 70   6. 25

For 7-12, find the greatest common factor for each pair of numbers.

7. 12 and 8   8. 48 and 60   9. 120 and 100

10. 45 and 50   11. 35 and 28   12. 90 and 72

For 13 - 15, complete the list of multiples of each number.

13. 3: 0, 3, 6, ☐, ☐, ☐   14. 9: 0 ☐, 18, ☐, ☐, ☐

15. 15: 0, ☐, ☐, ☐, ☐, 75

For 16-20, find the least common multiple of each pair of numbers.

16. 4 and 6   17. 12 and 15   18. 20 and 15

19. 12 and 4   20. 8 and 6

1.	
2.	
3.	
4.	
5.	
6.	
7.	
8.	
9.	
10.	
11.	
12.	
13.	
14.	
15.	
16.	
17.	
18.	
19.	
20.	

Review Exercises

1. Find the GCF of 40 and 56.

2. Find the LCM of 4 and 6

3. List all factors of 28.

4. List the first six multiples of 12.

5. Is $\frac{6}{8} = \frac{3}{4}$ a proportion? Why?

6. Solve the proportion.
$$\frac{6}{8} = \frac{n}{12}$$

Helpful Hints

Numbers can be assigned to a point on a **number line**. **Positive numbers** are to the right of zero. **Negative numbers** are to the left of zero.

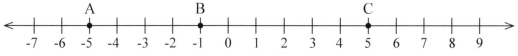

Numbers are graphed on a number line with a point.

Examples: A is the graph of -5.   A has a coordinate of -5.
B is the graph of -1.   B has a coordinate of -1.
C is the graph of 5.   C has a coordinate of 5.

Use the number line to state the coordinates of the given points.

S1. B

S2. D, E, and G

1. L and H

2. R and F

3. K, F, and C

4. N and A

5. G, H, I, and Q

6. H, D, and S

7. A, M, B, and P

8. B, C, and M

9. I, F, and P

10. L, P, H, and A

1.

2.

3.

4.

5.

6.

7.

8.

9.

10.

Problem Solving

At midnight the temperature was -4°.
By 6:00 a.m. the temperature had risen 12°.
What was the temperature at 6:00 a.m.?

Score

Review Exercises

Use A = {2,4,6,8,10}, B = {1,3,4,5,6,8,10}, and C = {4,5,6,8,9,10} to answer the following questions.

1. A ∩ B =   2. B ∪ C =   3. A ∪ C =   4. B ∩ C =

5. Are B and C equivalent sets? Why?   6. Are A and C disjoint sets? Why?

Helpful Hints

Equations can be solved and graphed on a **number line.**

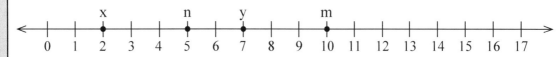

Examples:

$x + 5 = 7$	$n - 3 = 2$	$3y = 21$	$\frac{m}{2} = 5$
$2 + 5 = 7$	$5 - 3 = 2$	$3 \times 7 = 21$	$\frac{10}{2} = 5$
$x = 2$	$n = 5$	$y = 7$	
			$m = 10$

Solve each equation and graph each solution on the number line.
Also place each solution in the answer column.

S1. $x + 2 = 3$ S2. $y - 2 = 5$ 1. $c + 4 = 7$

2. $5 - e = 0$ 3. $3d = 15$ 4. $\frac{f}{3} = 6$

5. $n \times 2 = 8$ 6. $3 + j = 14$ 7. $3 + k = 11$

8. $4m = 28$ 9. $6 = n + 2$ 10. $\frac{r}{2} = 6$

1.	
2.	
3.	
4.	
5.	
6.	
7.	
8.	
9.	
10.	

Problem Solving

A car can travel 320 miles in five hours.
At this rate, how far can it travel in eight hours?

Score

Review Exercises

1. -2 + 9 =

2. -7 - 15 =

3. -7 - -15 =

4. 6 × -7 =

5. -45 ÷ -9 =

6. $\dfrac{-24 \div -2}{18 \div -3}$ =

Helpful Hints

Ordered pairs can be graphed on a **coordinate system**.

The first number of an ordered pair shows how to move across. It is called the **x-coordinate**.

The second number of an ordered pair shows how to move up and down. It is called the **y-coordinate**.

Examples: To locate B, move across to the right to 3 and up 4. The ordered pair is (3,4).

To locate C, move across to the left to -5 and up 2. The ordered pair is (-5,2).

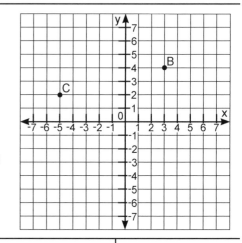

Use the coordinate system to find the ordered pair associated with each point.

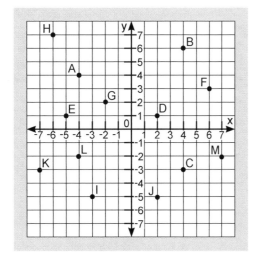

S1. D

S2. L

1. F

2. J

3. K

4. E

5. B

6. C

7. I

8. G

9. D

10. H

1.	
2.	
3.	
4.	
5.	
6.	
7.	
8.	
9.	
10.	
Score	

Problem Solving

A shirt that regularly sells for $30 is on sale for 20% off. How much is the sale price?

Review Exercises

1. $\dfrac{1}{3} + -\dfrac{4}{5} =$

2. $-.29 + -.39 =$

3. $-\dfrac{1}{8} - (-\dfrac{1}{2}) =$

4. $-\dfrac{2}{3} \times -1\dfrac{1}{2} =$

5. $2\dfrac{1}{2} \div -\dfrac{1}{2} =$

6. $-5 \div -2\dfrac{1}{2} =$

Helpful Hints

A **point** can be found by matching it with an ordered pair.

Examples: (-5, 3) is found by moving across to the left to -5, and up 3. This is represented by point B. -5 is the **x-coordinate** and 3 is the **y-coordinate**.

(6, 3) is found by moving across to the right to 6, and up 3. This is represented by point C. 6 is the **x-coordinate** and 3 is the **y-coordinate**.

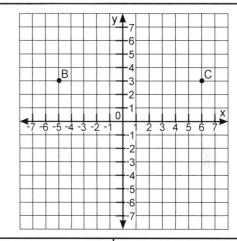

Use the coordinate system to find the point associated with each ordered pair.

S1. (6, 2)   S2. (-5, 5)

1. (3, 5)   2. (7, -6)

3. (-6, -4)   4. (0, 3)

5. (-2, -4)   6. (-2, 2)

7. (-6, 2)   8. (4, 2)

9. (-4, -7)   10. (4, -3)

1. _____
2. _____
3. _____
4. _____
5. _____
6. _____
7. _____
8. _____
9. _____
10. _____

Problem Solving

In a class of 40 students, 38 were present. What percent of the class was present?

Score _____

Review Exercises

1. $2^5 =$

2. $\sqrt{36} + 4^2 =$

3. $\dfrac{4^2 + 3^2}{\sqrt{25}} =$

4. $2^3 \times 3^2 =$

5. Write .00017 in scientific notation.

6. Write 213,000 in scientific notation.

Helpful Hints

The **slope** of a line refers to how steep the line is. It is the ratio of **rise to run**.

$$\text{slope} = \frac{y_2 - y_1}{x_2 - x_1}$$

Example:
What is the slope of the line passing through the ordered pairs (1, 5) and (6, 9)?

$$\text{slope} = \frac{y_2 - y_1}{x_2 - x_1} \qquad \begin{array}{cc} x_1 \; y_1 & x_2 \; y_2 \\ (1, 5), & (6, 9) \end{array}$$

$$= \frac{9 - 5}{6 - 1}$$

$$= \boxed{\frac{4}{5}} \qquad \text{The slope is } \frac{4}{5}$$

The run is 5 and the rise is 4.

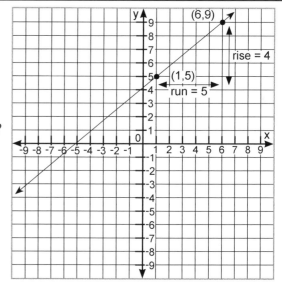

Find the slope of each line that passes through the given point.

S1. (2, 3), (5, 4)

S2. (3, -2), (5, 1)

1. (4, 3), (2, 6)

2. (4, 1), (7, 2)

3. (-2, 1), (-3, 3)

4. (-2, -2), (6, 3)

5. (4, 5), (6, 6)

6. (1, 2), (3, 9)

7. (1, -1), (6, 5)

8. (3, 2), (8, 6)

9. (2, -1), (4, 2)

10. (9, 2), (7, 5)

1.

2.

3.

4.

5.

6.

7.

8.

9.

10.

Problem Solving

In a school the ratio of boys to girls is five to four. If there are 400 boys, how many girls are there in the school?

Score

Reviewing Number Lines and Coordinate Systems

Use the number line to state the coordinates of the given points.

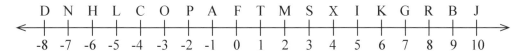

1. C   2. B, F, and J   3. S, M, and N   4. R, T, C, and D

Solve each equation and graph each solution on the number line. Be sure to label your answers. Also, place each solution in the answer column.

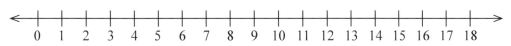

5. $n + 2 = 5$ 6. $x - 2 = 4$ 7. $3y = 15$ 8. $\frac{m}{2} = 4$

Use the coordinate system to find the ordered pair associated with each point.

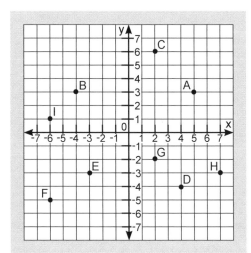

9. A

10. I

11. D

12. F

13. C

14. Find the slope of the line that passes through points A and G.

Use the coordinate system to find the ordered pair associated with each point.

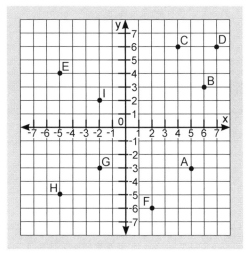

15. (6, 3)

16. (-2, 2)

17. (-5, -5)

18. (7, 6)

19. (5, -3)

20. Find the slope of the line that passes through points B and I.

1.	
2.	
3.	
4.	
5.	
6.	
7.	
8.	
9.	
10.	
11.	
12.	
13.	
14.	
15.	
16.	
17.	
18.	
19.	
20.	

Review Exercises

1. -36 ÷ -6 =

2. -9 -6 + -3 =

3. -2 × -3 × -4 =

4. -7 - 9 =

5. -56 ÷ 8 =

6. $(-2)^3$ =

Helpful Hints

The graph of a **linear equation** is always a line. A linear equation can have an infinite number of solutions, so to make a graph we select a few points and graph them, and then draw a line that connects them.

Example: Draw a graph of the solutions to the following equation.

$$y = x + 3$$

First, select four values for x and find the values for y. Start with x = 0 and make a chart like the one to the right.

x	y	
0	3	(0,3)
1	4	(1,4)
2	5	(2,5)
4	7	(4,7)

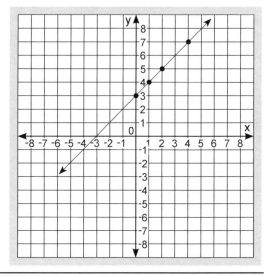

Next, plot the points and connect them with a line.

Make a table of 4 solutions. Graph the points. Connect them with a line

S1.

y = 2x + 1

x | y

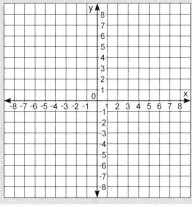

S2.

y = x - 3

x | y

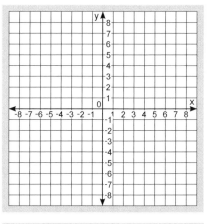

1.

y = 2x - 1

x | y

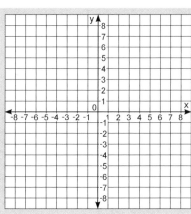

2.

$$y = \frac{x}{2}$$

x | y

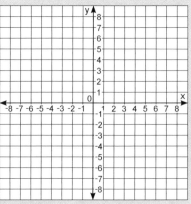

Problem Solving   If three pounds of meat costs $3.60, how much will five pounds costs?

Review Exercises

1. Solve the proportion.

$$\frac{5}{6} = \frac{7}{n}$$

2. Find 15% of 20.

3. 15 = 20% of what?

4. $-\frac{1}{3} + -\frac{3}{8} =$

5. $2 \times -1\frac{1}{2} =$

6. $-6.3 \div 3 =$

Helpful Hints

Use what you have learned to work the following problems.

Example: Draw a graph of the solutions to the following equation.

$$y = \frac{x}{2} + 2$$

x	y	
0	2	(0,2)
2	3	(2,3)
4	4	(4,4)
-2	1	(-2,1)

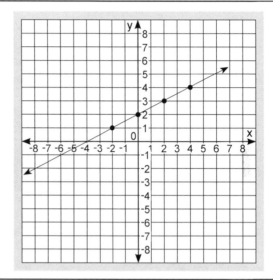

Make a table of 4 solutions. Graph the points. Connect them with a line

S1.

$$y = \frac{x}{3}$$

x	y

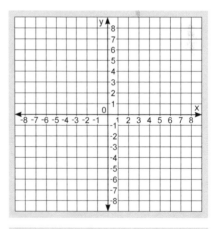

S2.

$$y = 2x + 3$$

x	y

1.

$$y = \frac{x}{2} + 5$$

x	y

2.

$$y = -2x$$

x	y

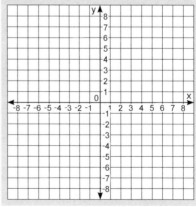

Problem Solving   John has finished $\frac{4}{5}$ of the problems on a test. What percent has he finished?

Review Exercises

1. Write 1,720,000 in scientific notation.

2. Write .00000038 in scientific notation.

3. Write 1.963×10^8 as a conventional number.

4. Write 3.4×10^{-4} as a conventional number.

5. $-9 - 7 - -6 =$

6. $-.34 + .53 =$

Helpful Hints

The goal with any **equation** is to end up with the **variable** (letter), an **equal sign**, and the **answer**. You can add, subtract, multiply, and divide on each side of the equal sign with the same number, and won't change the solution.

Examples:

$$x + 2 = 9$$
$$\underline{+\ -2 = -2}$$
$$x = \boxed{7}$$
Add -2 to both sides

$$n - 6 = -5$$
$$\underline{+\ 6 = 6}$$
$$n = \boxed{1}$$
Add 6 to both sides.

$$4n = 24$$
$$\frac{4n}{4} = \frac{24}{4}$$
$$n = \boxed{6}$$
Divide both sides by 4.

$$\frac{x}{6} = 4$$
$$\frac{6}{1} \times \frac{x}{6} = 4 \times 6$$
$$x = \boxed{24}$$
Multiply both sides by 6.

Check your work by substituting your answer in the original equation.

Solve the equations. Refer to the examples above.

S1. $x + 3 = 8$

S2. $3n = 96$

1. $n - 5 = -8$

2. $\frac{n}{5} = 8$

3. $n + 6 = -7$

4. $5n = -25$

5. $n + -6 = 7$

6. $\frac{n}{4} = -3$

7. $x + 23 = 57$

8. $15n = 60$

9. $n - -6 = -5$

10. $n + 12 = -15$

1.	
2.	
3.	
4.	
5.	
6.	
7.	
8.	
9.	
10.	
Score	

Problem Solving

What is the slope of a line that passes through the points (6, 1) and (9, 8)?

Review Exercises

1. $\dfrac{3^2 + 2^2 + 7}{2} =$

2. $3 + 7 \times 2 + 6 =$

3. $3 \times (7^2 - 15) =$

4. $n + 5 = -3$

5. $3n = 18$

6. $\dfrac{n}{3} = 7$

Helpful Hints

Be careful with negative signs when solving equations.

Examples:

$$\begin{aligned} -x + 7 &= -9 \\ + \ -7 &= -7 \\ \hline -x &= -16 \end{aligned}$$

If $-x = -16$,
then x = ⓰

$$\begin{aligned} -3n &= 18 \\ \dfrac{-3n}{-3} &= \dfrac{18}{-3} \end{aligned}$$

Divide both
sides by -3.

n = ⟨-6⟩

$$\dfrac{n}{-5} = 7$$

$$\dfrac{-5}{1} \times \dfrac{n}{-5} = 7 \times -5$$

Multiply both
sides by -5.

n = ⟨-35⟩

* Remember to
check your work
by substituting
your answer in the
original equation.

Solve the equations. Refer to the examples above.

S1. $-x + 7 = -5$

S2. $-4n = -12$

1. $-n - 6 = 8$

2. $\dfrac{n}{-2} = 6$

3. $-x + -7 = 2$

4. $-3n = 15$

5. $n - 6 = 12$

6. $5n = -30$

7. $-n - 6 = -8$

8. $\dfrac{n}{-4} = -5$

9. $3n = -45$

10. $-n + -6 = -20$

1.	
2.	
3.	
4.	
5.	
6.	
7.	
8.	
9.	
10.	

Problem Solving

Write the ratio 18 to 8 as a fraction reduced to lowest terms.

Score

Review Exercises

1. -7 - -9 + 6 - 7 =

2. 3 × -2 × 4 × -3 =

3. $\dfrac{-64 \div 8}{24 \div -6} =$

4. Write 210,000 in scientific notation.

5. Write .00316 in scientific notation.

6. $(-2)^4 =$

Helpful Hints

Some equations require two steps.

Examples:

$$2x - 5 = 71 \quad \text{Add 5 to}$$
$$+\ \ 5 =\ \ 5 \quad \text{both sides.}$$
$$\dfrac{2x}{2} = \dfrac{76}{2} \quad \text{Divide both sides by 2.}$$
$$x = \boxed{38}$$

$$\dfrac{n}{5} + 3 = 8 \quad \text{Add -3 to}$$
$$+\ -3 = -3 \quad \text{both sides.}$$
$$\dfrac{5}{1} \times \dfrac{n}{5} = 5 \times 5 \quad \text{Multiply both sides by 5.}$$
$$n = \boxed{25}$$

$$-3n - 4 = 11 \quad \text{Add 4 to}$$
$$+\ \ \ 4 =\ \ 4 \quad \text{both sides.}$$
$$\dfrac{-3n}{-3} = \dfrac{15}{-3} \quad \text{Divide both sides by -3.}$$
$$n = \boxed{-5}$$

* Remember to check your work by substituting your answer in the original equation.

S1. 3x - 5 = 16

S2. $\dfrac{x}{2} + 2 = 4$

1. 7x + 3 = -4

2. -14n - 7 = 49

3. 2n + 45 = 15

4. $\dfrac{n}{5} + -6 = 9$

5. 4x - 10 = 38

6. -2m + 9 = 7

7. 35x + 12 = 82

8. $\dfrac{m}{5} - 7 = 3$

9. 3x - 12 = 18

10. 5x + 2 = -13

1.

2.

3.

4.

5.

6.

7.

8.

9.

10.

Problem Solving

Six students were absent Monday at Jefferson School. If this was 3% of the total enrollment, how many students are enrolled at Jefferson School?

Score

Review Exercises

1. $-\frac{2}{5} + \frac{1}{2} =$

2. $-\frac{2}{5} + -\frac{2}{5} =$

3. $\frac{2}{3} \div -\frac{1}{2} =$

4. $1\frac{1}{2} \times -2 =$

5. $.2 \times -3.2 =$

6. $-.6 + -.5 =$

Helpful Hints

Sometimes the **distributive property** can be used to solve equations.

Examples:

$2(x + 7) = 30$
First use the distributive property.

$2(x + 7) = 30$
$2x + 14 = 30$
$\underline{+\quad -14 = -14}$ Add -14 to both sides.
$\frac{2x}{2} = \frac{16}{2}$ Divide both sides by 2.
$x = 8$

$3(4x - 3) = -33$
First use the distributive property.

$3(4x - 3) = -33$
$12x - 9 = -33$
$\underline{+\quad 9 = 9}$ Add 9 to both sides.
$\frac{12x}{12} = \frac{-24}{12}$ Divide both sides by 12.
$x = -2$

* Remember to check your answers.

Solve the following equations. Use the distributive property when necessary.

S1. $5(m + 6) = 45$

S2. $\frac{x}{5} + -6 = 3$

1. $3(m - 2) = 18$

2. $3x + 7 = -2$

3. $4m - 9 = 31$

4. $2(m + -2) = -10$

5. $-6x + 2 = -28$

6. $-x + 8 = 12$

7. $\frac{x}{2} + 3 = -2$

8. $2x + 1 = -13$

9. $5x - 3 = -18$

10. $4(x + 2) = 48$

1.	
2.	
3.	
4.	
5.	
6.	
7.	
8.	
9.	
10.	
Score	

Problem Solving

Find the greatest common factor of 42 and 56.

Review Exercises

1. $6^3 =$

2. $7^0 =$

3. $9^1 =$

4. $\sqrt{36} + \sqrt{49} =$

5. $2^3 + \sqrt{16}$

6. $33 + 5^3 =$

Helpful Hints

Sometimes there are variables on both sides of the equal sign.

Examples:

$$5x - 6 = 2x + 9$$
$$+ \ -2x \quad = -2x \quad \text{Add -2x to both sides.}$$
$$3x - 6 = 9$$
$$+ \quad 6 = 6 \quad \text{Add 6 to both sides.}$$
$$\frac{3x}{3} = \frac{15}{3} \quad \text{Divide both sides by 3.}$$
$$x = \boxed{5}$$

$$-6x + 12 = 4x - 8$$
$$+6x \quad = 6x \quad \text{Add 6x to both sides.}$$
$$12 = 10x - 8$$
$$+ \quad 8 = \quad 8 \quad \text{Add 8 to both sides.}$$
$$\frac{20}{10} = \frac{10x}{10} \quad \text{Divide both sides by 10.}$$
$$\boxed{2} = x$$

S1. $4x + 3 = 2x + 1$

S2. $5x + 1 = 2x + 10$

1. $4x - 12 = 2x + 2$

2. $x - 2 = 2x - 4$

3. $7x - 16 = x + 8$

4. $7x - 1 = 15 + 3x$

5. $-2x + 8 = 4x - 10$

6. $3x + 6 = x + 8$

7. $3x - 5 = x - 7$

8. $3x + 5 = x + 13$

9. $2x + 6 = -x + 12$

10. $5x - 6 = 2x + 12$

1.

2.

3.

4.

5.

6.

7.

8.

9.

10.

Problem Solving

Susan had 30 apples and used six of them to make a pie. What percent of the apples did she use to make the pie?

Score

Review Exercises

1. Solve the proportion.

   $\dfrac{n}{4} = \dfrac{25}{5}$

2. Is $\dfrac{4}{7} = \dfrac{3}{5}$ a proportion? Why?

3. Write 16 to 6 as a fraction reduced to lowest terms.

4. Find 20% of 300.

5. Six is what % of 24?

6. $7 = 20\%$ of what?

Helpful Hints

Use what you have learned to solve the following equations.
* If necessary, refer to the previous Helpful Hints sections.
* Check your answers by substituting them in the original equation.

Solve the following equations. Use the distributive property when necessary.

1.	
2.	
3.	
4.	
5.	
6.	
7.	
8.	
9.	
10.	
Score	

S1. $7n = 28$ S2. $\dfrac{n}{5} = 3$ 1. $2x + 6 = 2$

2. $3x + 7 = -2$ 3. $4m - 9 = 31$ 4. $4(x + 3) = -8$

5. $3(n - 2) = 30$ 6. $3x + 6 = x + 8$ 7. $4x - 12 = 2x + 2$

8. $-5x + 2 = -13$ 9. $\dfrac{x}{5} - 2 = -7$ 10. $3(x + 4) = -18$

Problem Solving

Jeff has a marble collection. The ratio of red marbles to blue marbles is three to two. If he has 12 red marbles, how many blue marbles does he have? (Use a proportion.)

Review Exercises

1. List all the factors of 48.

2. What is the GCF of 16 and 24?

3. What is the LCM of 6 and 10?

4. $3n = 15$, $n =$

5. $\dfrac{n}{2} = 10$, $n =$

6. $-x = 5$, $x =$

Helpful Hints

To solve **algebra word problems**, it is necessary to translate words into **algebraic expressions** containing a **variable**. A **variable** is a letter that represents a number. Here are some examples:

Three more than a number $\rightarrow x + 3$
Twice a number $\rightarrow 2x$
The quotient of x and five $\rightarrow \dfrac{x}{5}$
Seven less than three times a number $\rightarrow 3x - 7$
Twice a number less nine is equal to 15 $\rightarrow 2x - 9 = 15$
The difference between three times a number and eight equals 12 $\rightarrow 3x - 8 = 12$
The sum of a number and -9 is 24 $\rightarrow x + -9 = 24$
Three times a number less six equals twice the number plus 15 $\rightarrow 3x - 6 = 2x + 15$
Twice the sum of n and five $\rightarrow 2(n + 5)$
The difference between four times x and 15 equals twice the number $\rightarrow 4x - 15 = 2x$

Four less than a number $\rightarrow x - 4$
Seven times a number $\rightarrow 7x$
A number decreased by six $\rightarrow x - 6$

Translate each of the following into an equation.

S1. Seven less than twice a number is 12.

S2. Two more than three times a number equals 30.

1. The sum of twice a number and five is 14.

2. The difference between four times a number and six is 10.

3. Twelve is five less than four times a number.

4. One-third times a number less four equals twice the number added to eight.

5. Twice the sum of a number and two equals 10.

6. The difference between five times a number and three is 17.

7. Twice a number decreased by six is 15.

8. Two less than three times a number is seven more than twice the number.

9. Four more than a number equals the sum of seven and -12.

10. A number divided by five is 25.

1.	
2.	
3.	
4.	
5.	
6.	
7.	
8.	
9.	
10.	

Problem Solving

If a car can travel 65 miles per hour, how far can it travel in 3.5 hours?

Score

Review Exercises

1. x + 2 = 9
   x =

2. n + -3 = -7
   n =

3. 3n = 36

4. -5n = -25

5. $\frac{n}{3} = 5$
   n =

6. 2x + 1 = 7

Helpful Hints

Algebra word problems must be translated into an **equation** and solved.

Example:

Six times a number less two equals four times the number added to 10.
First translate and then solve.

$$6x - 2 = 4x + 10$$

$$\underline{+ \text{-}4x \qquad \text{-}4x}$$ Add -4x to both sides.

$$2x - 2 = 10$$

$$ 2 = 2$$ Add 2 to both sides.

$$\frac{2x}{2} = \frac{12}{2}$$ Divide both sides by 2.

$$x = \boxed{6}$$ The number is 6.

Translate each of the following into an equation and solve.

S1. Six less than twice a number is 16. Find the number.

S2. The difference between three times a number and 8 is 28.
Find the number.

1. Five less than twice a number is 67. Find the number.

2. Four times a number decreased by five is -17. Find the number.

3. Four times a number less six is eight more than two times the number.
Find the number.

4. Eight more then one-half a number is 10. Find the number.

5. The difference between four times a number and two is 10.

1.
2.
3.
4.
5.
Score

Problem Solving

A doctor's annual income is $150,000. What is his average monthly income?

Review Exercises

1. Write 3.61×10^{-7} as a conventional number.

2. Write .00000127 in scientific notation.

3. Write 729,000,000 in scientific notation.

Helpful Hints

Remember these steps when solving algebra word problems.

1. Read the problem very carefully.
2. Write an equation.
3. Solve the equation and find the answer.
4. Check your answer to be sure it makes sense.

Example: John is twice as old as Susan. The sum of their ages is 42. What is each of their ages?

Let x = Susan's age   2x = John's age

$x + 2x = 42$ Susan's age is x = (14.)
$3x = 42$ John's age is 2x = (28.)
$x = 14$ The sum is 42.

Solve the algebra word problems.

S1. Amir is six years older than Kevin. The sum of their ages is 30. Find the age of each.

S2. A board 44 inches long is cut into two pieces. The long piece is three times the length of the short piece. What is the length of each piece.

1. Bob and Bill together earn $66. Bill earned $6 more than twice as much as Bob. How much did each earn?

2. Steve worked Monday and Tuesday and earned a total of $212. He earned $30 more on Tuesday than he did on Monday. How much did Steve earn each day?

3. Five times Bob's age plus six equals three times his age plus 30. What is Bob's age?

4. Sixty dollars less than three times Susan's weekly salary is equal to 360 dollars. What is Susan's weekly salary?

5. Twice John's age less 12 is 48. What is John's age?

1.	
2.	
3.	
4.	
5.	
Score	

Problem Solving

A student has test scores of 90, 96, 84, and 86. What was his average score?

Review Exercises

Solve each equation.

1. $2x + 7 = -15$

2. $5x + 6 = 106$

3. $\frac{n}{4} + 2 = 13$

4. $3(n + 6) = -9$

5. $5x + 3 = 7x + -3$

6. $3x + 2x = 55$

Helpful Hints

*Remember: 1. Read the problem carefully.
2. Write an equation.
3. Solve the equation and find the answer.
4. Check your answer to be sure it makes sense.

Solve each algebra word problem.

S1. Five more than six times a number is equal to 48 less 7.
Find the number.

S2. Steve weighs 50 pounds more than Bart. Their combined weight is 270
pounds. What is each of their weights?

1. The sum of three times a number and 15 is -12. Find the number.

2. Eight more than six times a number is 20 more than four times the number.
Find the number.

3. The sum of five and a number is -19. Find the number.

4. Roy is three times as old as Ellen. The sum of their ages is 44 years.
What are each of their ages?

5. Six more than two times a number is six less than six times the number.
Find the number.

1.
2.
3.
4.
5.
Score

Problem Solving

A plane travelled 2,100 miles in 3.5 hours.
What was the plane's average speed per hour?

Reviewing Equations and Algebra Word Problems

For 1 - 12, solve each equation. Be sure to show all work.

1. $x + 5 = -2$

2. $3n = 39$

3. $\dfrac{n}{7} = 8$

4. $5n + 2 = 17$

5. $3n - 6 = -21$

6. $\dfrac{n}{3} - 6 = -12$

7. $3(n + 2) = -15$

8. $5(x - 4) = 55$

9. $2x + 4 = 4x - 12$

10. $5x - 3 = 3x + 13$

11. $3x + 4x = -77$

12. $\dfrac{n}{-3} + 2 = -5$

For 13 - 20, solve each algebra word problem

13. Twice a number less three is 21. Find the number.

14. Eight more than five times a number is -17. Find the number.

15. The difference between five times a number and six is 24. Find the number.

16. Seven more than twice a number is five less than four times the number. Find the number.

17. Ann has twice as much money as Sue. Together they have $66. How much does each have?

18. Bill is eight years older than Ron. The sum of their ages is 64 years. How old is each of them?

19. Four times a number decreased by six equals -14. Find the number.

20. Four more than one-third of a number is 10. Find the number.

1.	
2.	
3.	
4.	
5.	
6.	
7.	
8.	
9.	
10.	
11.	
12.	
13.	
14.	
15.	
16.	
17.	
18.	
19.	
20.	

Review Exercises

1. List the first seven multiples of 8.

2. List all factors of 60.

3. What is the GCF of 100 and 40?

4. Write .000006 in scientific notation.

5. Write 2,100,000 in scientific notation.

6. Write 2.1×10^{-3} as a conventional number.

Helpful Hints

Probability tells what chance, or how likely it is for an event to occur. Probability can be written as a fraction.

$$\textbf{Probability} = \frac{\text{number of ways a certain outcome can occur}}{\text{number of possible outcomes}}$$

Examples: If you toss a coin, what is the probability that it will show heads?

$$\frac{1 - \text{heads is one outcome}}{2 - \text{there are two possible outcomes, heads or tails}}$$ The probability is 1 out of 2.

There are six marbles in a jar. Three are red, two are blue, and one is green. What is the probability that you will draw a blue one without looking?

$$\frac{2 - \text{blue marbles}}{6 - \text{marbles in the jar}}$$ The probability is 2 out of 6, or simplified, 1 out of 3.

Use the information below to answer the following questions.

There are 3 red marbles, 6 blue marbles, 2 black marbles, and 1 green marble in a can. Find the probability of each of the following.

S1. A red marble.

S2. A blue or green marble.

1. A black marble.

2. A green marble.

3. A blue or red marble.

4. Not a black marble.

5. Not a red marble.

6. Not a green or blue marble.

7. A green, red, or blue marble.

8. Not a blue marble.

9. A green, red, or black marble.

10. Not a blue or black marble.

1.	
2.	
3.	
4.	
5.	
6.	
7.	
8.	
9.	
10.	

Problem Solving

Four times a number less five is -17. Find the number.

Score

Review Exercises

Solve each of the following equations.

1. $3x + 2 = -28$

2. $\frac{x}{5} - 6 = -11$

3. $4(n + 3) = -28$

4. $2x + 10 = 4x + 2$

5. $3x + 2x = 75$

6. $7x - 3 = 60$

Helpful Hints

Use what you have learned to solve the following questions.

Example: What is the probability of the spinner landing on the 1 or the 3?

2 out of 8 or, simplified, 1 out of 4.

Use the spinner to find the probability for each of the following.
Find the probability of spinning once and landing on each of the following.

S1. a three

S2. an even number

1. a seven

2. not a five

3. an odd number

4. a number less than five

5. a number greater than six

6. a nine

7. a one or an eight

8. an even number or a five

9. a number greater than three

10. a number which is a factor of six

1.
2.
3.
4.
5.
6.
7.
8.
9.
10.
Score

Problem Solving

If five pounds of beef cost $9, how many pounds can be bought with $36?

Review Exercises

1. Change .3 to a percent.   2. Change .03 to a percent.   3. Change $\frac{3}{5}$ to a percent.

4. Find 4% of 50.   5. Fifteen is what % of 60?   6. 4 = 20% of what?

Helpful Hints

Statistics involves gathering and recording **data**. Number facts about events or objects are called data. The **range** is the difference between the greatest number and the least number in a list of data. The **mode** is the number which appears the most in a list of data.

Example: Find the range and mode for the list of data.
12, 10, 1, 7, 4, 7, 5
First, list the numbers from least to greatest.
1, 4, 5, 7, 7, 10, 12
The range is 12 - 1 = 11.
The mode is 7, which appears the most.

Arrange the data in order from least to greatest, then find the range and mode.

S1. 7, 4, 1, 8, 2, 5, 4

S2. 6, 2, 7, 6, 8, 2, 5, 6, 3

1. 7, 4, 8, 2, 4, 7, 7

2. 25, 17, 30, 39, 16, 24, 30

3. 1, 3, 6, 3, 4, 6, 11, 9

4. 1, 6, 17, 8, 9, 20, 9

5. 7, 3, 1, 3, 1, 3, 8, 4

6. 3, 14, 8, 6, 11, 8, 14, 8

7. 1, 10, 2, 9, 3, 8, 2, 7

8. 85, 91, 90, 86, 91, 87

9. 1, 10, 2 9, 2, 7, 2, 8

10. 20, 2, 19, 1, 2, 16, 3

1.	
2.	
3.	
4.	
5.	
6.	
7.	
8.	
9.	
10.	
Score	

Problem Solving

If three cans of juice cost $1.14, what is the cost of one can?

Review Exercises

1. Write 16 to 10 as a fraction reduced to lowest terms.

2. Is $\frac{9}{11} = \frac{7}{8}$ a proportion? Why?

3. Solve the proportion.
$$\frac{4}{n} = \frac{9}{45}$$

4. Write 1,280,000 in scientific notation.

5. Write .0000962 in scientific notation.

6. Write 6.2×10^{-5} as a conventional number.

Helpful Hints

The **mean** of a list of data is found by adding all the items in the list and then dividing by the number of items.

The **median** is the middle number, when the list of data is arranged from least to greatest.

Example: Find the mean and median for the list of data.

1, 2, 5, 6, 6

Median = ⑤   Mean = $\dfrac{1 + 2 + 5 + 6 + 6}{5} = \dfrac{20}{5} = ④$

Arrange the data in order from least to greatest, then find the mean and median.

S1. 1, 5, 2, 4, 3

S2. 6, 1, 7, 4, 2, 6, 2

1. 2, 7, 1, 4, 1

2. 1, 5, 7, 1, 2, 2, 3

3. 5, 25, 10, 20, 15

4. 1, 1, 1, 3, 3, 3, 4, 1, 1

5. 8, 5, 2, 9, 3, 6, 9

6. 126, 136, 110

7. 7, 3, 4, 2, 4

8. 3, 1, 4, 7, 5

9. 2, 10, 4, 8, 1

10. 50, 70, 30

1.
2.
3.
4.
5.
6.
7.
8.
9.
10.
Score

Problem Solving

In a class of 40 students, 20% of them received A's. How many students did not receive A's?

Review Exercises

1. $3 + 4 \times 5 - 2 =$

2. $3(8 + 2) - 4^2 =$

3. $(15 - 8) + 64 \div 2^3 =$

4. $7 \times 4 - 39 \div 3 =$

5. $6\,[(3 + 4) \times 2] =$

6. $3(-2 + 4) + 5 =$

Helpful Hints

Use what you have learned to answer the following questions.

* If necessary, refer to the two previous pages.

Arrange the data in order from least to greatest, then answer the questions.

2, 8, 6, 2, 7

S1. What is the range?   S2. What is the mode?

1. What is the mean?   2. What is the median?

1, 9, 2, 7, 2, 3, 4

3. What is the median?   4. What is the mode?

5. What is the range?   6. What is the mean?

2, 11, 8, 6, 1, 2, 5

7. What is the range?   8. What is the mode?

9. What is the mean?   10. What is the median?

1.
2.
3.
4.
5.
6.
7.
8.
9.
10.
Score

Problem Solving

Light travels at a speed of 1.86×10^5 miles per second.
Write the speed as a conventional number.

Reviewing Probability and Statistics

There are four green marbles, three red marbles, two white marbles, and one blue marble in a can. What is the probability for each of the following?

1. a red marble

2. a green marble

3. a green or blue marble

4. not a red marble

5. a green, red, or blue marble

6. not a green marble

Use the spinner to find the probability of spinning once and landing on each of the following.

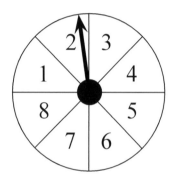

7. a five

8. an odd number

9. a number greater than three

10. a one or a three

11. a number less than five

12. a one or a six

Arrange the data in order from least to greatest, then answer the questions.

2, 5, 4, 10, 4

13. What is the range?

14. What is the mode?

15. What is the mean?

16. What is the median?

2, 5, 2, 1, 3, 7, 8

17. What is the mode?

18. What is the mean?

19. What is the range?

20. What is the median?

1.
2.
3.
4.
5.
6.
7.
8.
9.
10.
11.
12.
13.
14.
15.
16.
17.
18.
19.
20.

Practice Problems – Adding Integers

Add each of the following:

1) -26 + 19   2) -18 + 25

3) 76 + -87   4) -39 + -62

5) -96 + 103   6) -626 + 342

7) -63 + -97   8) -349 + -632

9) -632 + -908   10) -346 + 800

11) -96 + 13 + -12

12) 37 + -18 + 15 + -19

13) -49 + -16 + 92 + -16

14) 363 + -95 + -726

15) -37 + 23 + -16 + 530

1)	
2)	
3)	
4)	
5)	
6)	
7)	
8)	
9)	
10)	
11)	
12)	
13)	
14)	
15)	
Score:	

Practice Problems – Subtracting Integers

Subtract each of the following.

1) -7 – 9	2) 7 – 12
3) 16 – 19	4) 5 – -6
5) -17 – -26	6) -17 – 18
7) 33 – -16	8) -96 – 72
9) 43 – 16	10) -12 – -18
11) 3 – 7 – -6	12) 72 – 96
13) -99 – 103	14) -76 – 96
15) -363 – -900	16) 43 – -63
17) -42 – -43	18) -302 – 643
19) -4 – 96	20) -96 – -132

1)

2)

3)

4)

5)

6)

7)

8)

9)

10)

11)

12)

13)

14)

15)

16)

17)

18)

19)

20)

Score:

Practice Problems – Multiplying Integers

Multiply each of the following.

1) 7 x -33   2) -92 • 3

3) -8 • -19   4) -23(16)

5) (-32)(-12)   6) -7 x -63

7) 9(-32)   8) -24 • 25

9) (-26)(12)   10) -6 x -102

11) (-2)(-3)(-6)   12) 3(-6)(-4)

13) 3(-5) • 4(-6)   14) -4(-3) • 3(-5)

15) (2)(6)(-2)(-3)   16) 12(-9)(-2)

17) (-1)(-2)(-3)(-4)   18) 7 • (-6)(-4)

19) (-4)(3)(-2)(-1)   20) (63)(-2)(4)

1) _____

2) _____

3) _____

4) _____

5) _____

6) _____

7) _____

8) _____

9) _____

10) _____

11) _____

12) _____

13) _____

14) _____

15) _____

16) _____

17) _____

18) _____

19) _____

20) _____

Score: _____

Practice Problems – Dividing Integers

Divide each of the following.

1) $\dfrac{-48}{6}$

2) $\dfrac{-80}{-5}$

3) $-72 \div 4$

4) $-384 \div 6$

5) $-340 \div -2$

6) $\dfrac{168}{-12}$

7) $\dfrac{-225}{25}$

8) $\dfrac{368}{-16}$

9) $\dfrac{64 \div -8}{12 \div -3}$

10) $\dfrac{4 \bullet -12}{-10 \div 5}$

11) $\dfrac{6 \bullet (-8)}{2 \bullet (-1)}$

12) $\dfrac{-6\,(-16)}{-12 \div -6}$

13) $\dfrac{-45 \div -5}{-15 \div -5}$

14) $\dfrac{5\,(-60)}{-20 \div -2}$

15) $\dfrac{-42 \div 2}{-7 \bullet -3}$

16) $\dfrac{6 \bullet -12}{3\,(-8)}$

17) $\dfrac{-81 \div -9}{-27 \div 3}$

18) $\dfrac{5\,(-15)}{20 \div -4}$

19) $\dfrac{-84 \div 2}{-24 \div 4}$

20) $\dfrac{(-6)\,(-3)\,(2)}{(-2)\,(9)}$

1) _____

2) _____

3) _____

4) _____

5) _____

6) _____

7) _____

8) _____

9) _____

10) _____

11) _____

12) _____

13) _____

14) _____

15) _____

16) _____

17) _____

18) _____

19) _____

20) _____

Score: _____

Practice Problems – Positive and Negative Fractions

Simplify each of the following.

1) $-\frac{1}{3} + \frac{1}{2}$

2) $\frac{3}{4} + -\frac{2}{5}$

3) $-\frac{3}{4} + -\frac{1}{3}$

4) $3\frac{1}{2} + -1\frac{1}{4}$

5) $\frac{1}{3} - \frac{2}{3}$

6) $\frac{2}{5} - \frac{3}{4}$

7) $-\frac{2}{5} - \frac{1}{2}$

8) $\frac{1}{2} + -\frac{1}{3}$

9) $\frac{1}{4} - \frac{2}{5}$

10) $-\frac{4}{5} + -\frac{1}{3}$

11) $-\frac{3}{4} \times -1\frac{1}{2}$

12) $2\frac{1}{2} \cdot -\frac{1}{2}$

13) $\frac{5}{6} \times -\frac{3}{10}$

14) $-2\frac{1}{3} \div -1\frac{1}{7}$

15) $-3 \times 1\frac{1}{2}$

16) $-2\frac{1}{2} \div -1\frac{1}{4}$

17) $-3\frac{1}{2} \div \frac{1}{2}$

18) $-\frac{3}{4} \div -\frac{1}{3}$

19) $\frac{3}{5} + -\frac{1}{2}$

20) $-\frac{2}{3} + -\frac{1}{4}$

1) _____
2) _____
3) _____
4) _____
5) _____
6) _____
7) _____
8) _____
9) _____
10) _____
11) _____
12) _____
13) _____
14) _____
15) _____
16) _____
17) _____
18) _____
19) _____
20) _____

Score: _____

Practice Problems – Positive and Negative Decimals

Simplify each of the following.

1) -7.26 + 4.9

2) -6.73 + -5.96

3) 3.61 + -2.33

4) -2.6 + 7.9562

5) -3.9 + -4.37

6) 3.1 – 4.63

7) -21.6 – 7.68

8) 3 – 5.67

9) -.7 – .9

10) 3.632 – 7.97

11) 5 x -2.6

12) -3.36 • -.4

13) -2.3 x -4.6

14) 1.5 x -2.13

15) 5.32 x -1.2

16) -6.22 ÷ 2

17) -8.112 ÷ .3

18) -5.25 ÷ -5

19) -3.148 ÷ .4

20) -6.25 ÷ -.5

1) _____
2) _____
3) _____
4) _____
5) _____
6) _____
7) _____
8) _____
9) _____
10) _____
11) _____
12) _____
13) _____
14) _____
15) _____
16) _____
17) _____
18) _____
19) _____
20) _____

Score: _____

Practice Problems – Exponents

For 1-10, rewrite each as an integer.

1) 9^2

2) $(-3)^2$

3) $(-6)^3$

4) 5^3

5) $(-7)^4$

6) 3^4

7) $(-8)^3$

8) $(-9)^4$

9) 2^8

10) $(-7)^3$

For 11-20, rewrite each as an exponent.

11) $5 \times 5 \times 5 \times 5 \times 5$

12) 36

13) $(-2)(-2)(-2)(-2)$

14) 144

15) 225

16) $(-9)(-9)(-9)$

17) 100

18) 2,500

19) $(-10)(-10)(-10)(-10)$

20) 49

1) _____

2) _____

3) _____

4) _____

5) _____

6) _____

7) _____

8) _____

9) _____

10) _____

11) _____

12) _____

13) _____

14) _____

15) _____

16) _____

17) _____

18) _____

19) _____

20) _____

Score: _____

Practice Problems – Laws of Exponents

Simplify each of the following.

1) $6^3 \times 6^5$ 2) $(\frac{8}{5})^3$

3) $(4 \times 6)^3$ 4) $(5^3)^4$

5) $\frac{3^7}{3^2}$ 6) 6^{-4}

7) $7^3 \cdot 7^5$ 8) $(3 \cdot 9)^3$

9) $(4^2)^5$ 10) 3^{-3}

11) 6^1 12) 12^0

13) $\frac{4^2}{4^4}$ 14) $(5^3)^3$

15) 5^{-3} 16) $(\frac{7}{8})^4$

17) $\frac{3^8}{3^2}$ 18) $\frac{5^2}{5^5}$

19) $5^3 \cdot 5^2 \cdot 5^4$ 20) $\frac{6^5}{6^2}$

1) _____

2) _____

3) _____

4) _____

5) _____

6) _____

7) _____

8) _____

9) _____

10) _____

11) _____

12) _____

13) _____

14) _____

15) _____

16) _____

17) _____

18) _____

19) _____

20) _____

Score: _____

Practice Problems – Square Roots

Simplify each of the following.

1) $\sqrt{36}$

2) $\sqrt{121}$

3) $\sqrt{2500}$

4) $\sqrt{400}$

5) $\sqrt{75}$

6) $\sqrt{8}$

7) $\sqrt{\dfrac{144}{9}}$

8) $\sqrt{3600}$

9) $\sqrt{\dfrac{20}{36}}$

10) $\sqrt{\dfrac{12}{25}}$

11) $\sqrt{50}$

12) $\sqrt{72}$

13) $\sqrt{12}$

14) $\sqrt{200}$

15) $\sqrt{\dfrac{16}{4}}$

16) $\sqrt{27}$

17) $\sqrt{125}$

18) $\sqrt{4900}$

19) $\sqrt{\dfrac{72}{8}}$

20) $\sqrt{\dfrac{25}{36}}$

1) _____

2) _____

3) _____

4) _____

5) _____

6) _____

7) _____

8) _____

9) _____

10) _____

11) _____

12) _____

13) _____

14) _____

15) _____

16) _____

17) _____

18) _____

19) _____

20) _____

Score: _____

Practice Problems – Order of Operations

Solve each of the following.
Be sure to follow the correct order of operations.

1) $6 + 7 \times 4 - 3$

2) $9 + 2^3 \times 4 - 3$

3) $6^2 (24 \div 6) \div 2$

4) $7(8 + 4) - 6^2$

5) $7 + \{(7 \times 5) + 3\}$

6) $5(6 + -11) + -8$

7) $5^3 - 6(4 + 3)$

8) $(16 + -7) + 72 \div 3^2$

9) $\dfrac{(16 - 8) + 4^2}{-10 + 3(4 + 2)}$

10) $\dfrac{6^2 + (-9 + 5)}{3(5^2 - 15) - 14}$

11) $7(6 + 7) - 5^2$

12) $5^2(10 + 3) + -8$

13) $15 \div 5 \cdot 4 - 6$

14) $5\{(6 + 2) \cdot 4 - 5\} \div 3$

15) $2(6^2 - 5) - 12$

16) $3 + \{4(8 - 2) \div 6\} - 5$

17) $\dfrac{5^2(4^2)}{2(5 - 1)}$

18) $3^3 + 5(3 - 9)$

19) $12 + 4 \cdot 6 \div 2 + 5$

20) $9 + 4^2 \times 3 - 12$

1) _____
2) _____
3) _____
4) _____
5) _____
6) _____
7) _____
8) _____
9) _____
10) _____
11) _____
12) _____
13) _____
14) _____
15) _____
16) _____
17) _____
18) _____
19) _____
20) _____

Score: _____

Practice Problems – Properties of Numbers

Name the property that is illustrated.

1) $a \times (b + c) = a \times b + a \times c$　　　　2) $0 + a = a$

3) $a + b = b + a$　　　　4) $a \times b = b \times a$

5) $a + {-a} = 0$　　　　6) $a \times \frac{1}{a} = 1$　　$a \neq 0$

7) $(a + b) + c = a + (b + c)$　　　　8) $1 \times a = a$

9) $19 + {-19} = 0$　　　　10) $0 + (-12) = -12$

11) $15 \times 7 = 7 \times 15$　　　　12) $1 \times \frac{3}{4} = \frac{3}{4}$

13) $3(7) + 3(8) = 3(7 + 8)$　　　　14) $62 + 37 = 37 + 62$

15) $3 \times (7 \times 2) = (3 \times 7) \times 2$　　　　16) $c + {-c} = 0$

17) $a(b + c) = a(b) + a(c)$　　　　18) $(2 \times 3)\,2 = (3 \times 2)\,2$

19) $m + {-m} = 0$　　　　20) $(a + b) + c = (b + a) + c$

1)	_____
2)	_____
3)	_____
4)	_____
5)	_____
6)	_____
7)	_____
8)	_____
9)	_____
10)	_____
11)	_____
12)	_____
13)	_____
14)	_____
15)	_____
16)	_____
17)	_____
18)	_____
19)	_____
20)	_____
Score:	

Practice Problems – The Number Line

Use the number line to state the coordinates of the given points.

1) B   2) D, E, and K

3) L, S, and P   4) A, P, Q, and M

5) N and A   6) B, C, L, and G

7) A, J, D, and B   8) L, P, H, and A

9) R, P, B, and F

For 10-20, write the numbers in order from least to greatest.
Remember, lower values to the left, and higher values to the right.

10) -15, 20, 5   11) -5, $\frac{1}{5}$, 15

12) $\frac{3}{5}$, $\frac{7}{5}$, 1   13) $\frac{5}{4}$, $\frac{5}{1}$, $\frac{4}{5}$

14) 2^3, 5^2, 3^3 15) $3\frac{1}{3}$, $4\frac{1}{2}$, $2\frac{1}{8}$

16) $\frac{8}{5}$, 1, $\frac{3}{4}$ 17) -103, 100, 180

18) $\frac{1}{2}$, $\frac{3}{4}$, $\frac{1}{4}$ 19) $3\frac{1}{2}$, 2, $1\frac{7}{8}$

20) -7, 9, -15

1)
2)
3)
4)
5)
6)
7)
8)
9)
10)
11)
12)
13)
14)
15)
16)
17)
18)
19)
20)
Score:

Practice Problems – The Coordinate Plane

Use the coordinate system to find the ordered pair associated with each point.

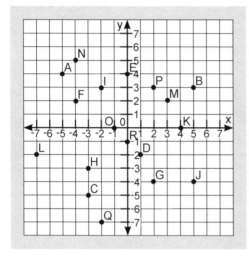

1) A	2) B
3) C	4) D
5) E	6) F
7) G	8) H
9) I	10) J

Use the coordinate system to find the point associated with each ordered pair.

11) (2, 3)	12) (3, 2)
13) (-1, 0)	14) (0, -1)
15) (5, -4)	16) (-4, 5)
17) (-4, 2)	18) (-2, -7)
19) (5, 3)	20) (-7, -2)

1) _____

2) _____

3) _____

4) _____

5) _____

6) _____

7) _____

8) _____

9) _____

10) _____

11) _____

12) _____

13) _____

14) _____

15) _____

16) _____

17) _____

18) _____

19) _____

20) _____

Score: _____

Practice Problems – Relations and Functions

1) Is M = {(2,3), (4,5), (7,8), (7,9)} a function? Why?

2) Is F = {(1,3), (4,7), (3,5)} a function? Why?

3) What is the domain of M?

4) What is the range of M?

5) What is the domain of F?

6) What is the range of F?

7) Explain the vertical line test.

8) All relations are functions. True or False?

9) All functions are relations. True or False?

Use F = {(1,2), (3,5), (4,5), (6,9)} to answer 10-12.

10) List the domain of F.

11) List the range of F.

12) Is F a function? Why?

Use R = {(3,4), (5,6), (5,8), (7,9)} to answer 13-15.

13) List the domain of R.

14) List the range of R.

15) Is R a function? Why?

1) _____

2) _____

3) _____

4) _____

5) _____

6) _____

7) _____

8) _____

9) _____

10) _____

11) _____

12) _____

13) _____

14) _____

15) _____

Score: _____

Practice Problems – Factors, Divisibility Tests & Prime Factorization

Use shortcut division to find the prime factorization of each of the following. Express your answers using exponents.

1) 42

2) 36

3) 54

4) 310

5) 125

6) 642

7) 93

8) 98

9) 105

10) 96

11) 144

12) 200

13) 135

14) 216

15) 60

16) 240

17) 127

18) 90

19) 102

20) 180

1) _____

2) _____

3) _____

4) _____

5) _____

6) _____

7) _____

8) _____

9) _____

10) _____

11) _____

12) _____

13) _____

14) _____

15) _____

16) _____

17) _____

18) _____

19) _____

20) _____

Score: _____

Practice Problems – Greatest Common Factors

Find the greatest common factor for each of the following.

1) 10, 15   2) 12, 28

3) 14, 35   4) 18, 24

5) 75, 50   6) 66, 90

7) 70, 108   8) 144, 200

9) 96, 156   10) 22, 33

11) 90, 135   12) 28, 98

13) 15, 24, 36   14) 132, 242

15) 24, 36, 30   16) 105, 210

17) 98, 154   18) 348, 426

19) 105, 126, 210   20) 90, 126, 252

1) _____
2) _____
3) _____
4) _____
5) _____
6) _____
7) _____
8) _____
9) _____
10) _____
11) _____
12) _____
13) _____
14) _____
15) _____
16) _____
17) _____
18) _____
19) _____
20) _____

Score: _____

Practice Problems – Least Common Multiples

Find the least common multiple for each of the following.

1) 12, 15   2) 20, 15

3) 22, 33   4) 14, 70

5) 16, 18   6) 6, 35

7) 24, 20   8) 45, 75

9) 28, 98   10) 15, 22

11) 18, 12   12) 60, 90

13) 5, 6, 10   14) 6, 8, 12

15) 15, 10, 20   16) 30, 24, 40

17) 120, 200   18) 48, 40

19) 100, 56   20) 72, 80

1) _____

2) _____

3) _____

4) _____

5) _____

6) _____

7) _____

8) _____

9) _____

10) _____

11) _____

12) _____

13) _____

14) _____

15) _____

16) _____

17) _____

18) _____

19) _____

20) _____

Score: _____

Practice Problems – Scientific Notation

For 1-10, change each of the following to scientific notation.

1) 14,490,000,000   2) .0000346

3) 259,790   4) .00000721

5) 1,079,000,000   6) .0076

7) .00000019   8) 240,000

9) .0000762   10) 54,000,000

For 11-20, change each number in scientific notation to a conventional number.

11) 6.093×10^5 12) 2.3×10^5

13) 6×10^{-6} 14) 1.347×10^4

15) 3.21×10^5 16) 4.2×10^{-3}

17) 4.5×10^5 18) 5×10^6

19) 3.72×10^{-3} 20) 3.95×10^6

1) _____

2) _____

3) _____

4) _____

5) _____

6) _____

7) _____

8) _____

9) _____

10) _____

11) _____

12) _____

13) _____

14) _____

15) _____

16) _____

17) _____

18) _____

19) _____

20) _____

Score: _____

Practice Problems – Ratios and Proportions

Solve each of the following proportions.

1) $\dfrac{x}{5} = \dfrac{3}{4}$ 2) $\dfrac{y}{3} = \dfrac{4}{7}$

3) $\dfrac{5}{x} = \dfrac{3}{2}$ 4) $\dfrac{x}{5} = \dfrac{3}{20}$

5) $\dfrac{4}{x} = \dfrac{12}{60}$ 6) $\dfrac{16}{x} = \dfrac{12}{9}$

7) $\dfrac{9}{x} = \dfrac{2}{3}$ 8) $\dfrac{2}{y} = \dfrac{5}{9}$

9) $\dfrac{8}{5} = \dfrac{10}{x}$ 10) $\dfrac{3}{8} = \dfrac{12}{x}$

11) $\dfrac{5}{x} = \dfrac{9}{11}$ 12) $\dfrac{2}{7} = \dfrac{5}{x}$

13) $\dfrac{18}{x} = \dfrac{2}{7}$ 14) $\dfrac{4}{9} = \dfrac{7}{x}$

15) $\dfrac{100}{x} = \dfrac{90}{45}$ 16) $\dfrac{65}{10} = \dfrac{13}{x}$

17) $\dfrac{x}{63} = \dfrac{2}{9}$ 18) $\dfrac{4}{5} = \dfrac{28}{x}$

19) $\dfrac{x}{5} = \dfrac{3}{7}$ 20) $\dfrac{8}{9} = \dfrac{x}{7}$

1) _____
2) _____
3) _____
4) _____
5) _____
6) _____
7) _____
8) _____
9) _____
10) _____
11) _____
12) _____
13) _____
14) _____
15) _____
16) _____
17) _____
18) _____
19) _____
20) _____

Score: _____

Practice Problems – Using Proportions in Word Problems

Use a proportion to solve each of the following.

1) If 3 tickets to a show cost $66.00, find the cost of 7 tickets.

2) If 3 cans of peas cost $1.65, then how many cans can be bought of $2.75?

3) A man can travel 11 miles on his bike in 2 hours. At this same rate, how far can he travel in 5 hours?

4) A train can travel 90 miles in $1\frac{1}{2}$ hours. At this rate, how far will the train travel in 6 hours?

5) A car can travel 150 km on 12 liters of gas. How many liters of gas are needed to travel 500 km?

6) Four cups of sugar are used for every 5 cups of flour in a recipe. How many cups of flour are needed for 64 cups of sugar?

7) How much would you pay for 5 apples if a dozen cost $4.80?

8) A recipe calls for $1\frac{1}{2}$ cups of sugar for a 3-pound cake. How many cups of sugar should be used for a 5-pound cake?

9) The scale on a map is 1 inch = 500 miles. If two cities are 875 miles apart, how far apart are they on this map?

10) A man received $420 for working 20 hours. At the same rate of pay, how many hours must he work to earn $735?

11) In a school, the ratio of the number of boys to the number of girls is 5 to 4. If 560 girls attend the school, what is the number of boys attending the school?

12) If 6 apples cost $.99, then how much will ten apples cost?

13) If a watch loses 2 minutes every 15 hours, then how much time will it lose in 2 hours?

14) A picture $3\frac{1}{4}$ inches long and $2\frac{1}{8}$ inches wide is to be enlarged so that the length will become $6\frac{1}{2}$ inches. What will be the width of the enlarged picture?

15) To mix concrete, the ratio of cement to sand is 1 to 4. How many bags of cement would be used with 100 bags of sand?

1)	
2)	
3)	
4)	
5)	
6)	
7)	
8)	
9)	
10)	
11)	
12)	
13)	
14)	
15)	
Score:	

Practice Problems – Percents

Solve each of the following.

1) Find 20% of 210.

2) Find 6% of 350.

3) 15 is what % of 60?

4) 5 = 20% of what?

5) 15 = 75% of what?

6) 30% of 200 =   .

7) 18 is what % of 24?

8) Find 25% of 64.

9) 3 is what % of 12?

10) 16 is what % of 80?

11) 3 is what % of 60?

12) Find 8% of 320.

13) 20 = 25% of what?

14) Find 40% of 60.

15) 12 = 50% of what?

16) 60 = what % of 80?

17) 3 is 50% of what?

18) Find 100% of 320.

19) 19 is what % of 76?

20) Find 45% of 450.

1) _____

2) _____

3) _____

4) _____

5) _____

6) _____

7) _____

8) _____

9) _____

10) _____

11) _____

12) _____

13) _____

14) _____

15) _____

16) _____

17) _____

18) _____

19) _____

20) _____

Score:

Practice Problems – Percent Word Problems

Solve each of the following.

1) Find 20% of 150.

2) 6 is 20% of what?

3) 8 is what % of 40?

4) Change $\frac{18}{20}$ to a percent.

5) A school has 600 students. If 4% are absent, how many students are absent?

6) A quarterback threw 24 passes and 75% were caught. How many were caught?

7) Riley has 250 marbles in his collection. If 50 of them are red, what percent of them are red?

8) A team played 60 games and won 45 of them. What percent did they win?

9) There are 50 sixth graders in a school. This is 20% of the school. How many students are in the school total?

10) A coat is on sale for $20. This is 80% of the regular price. What is the regular price?

11) Steve has finished $\frac{3}{5}$ of his test. What percent of the test has he finished?

12) Alex wants to buy a computer priced at $640. If sales tax is 8%, what is the total cost of the computer?

13) On a test with 120 problems, a student got 90 correct. What percent were correct?

14) A man bought a car for $11,700. If the sales tax was 8%, how much was the sales tax?

15) At a school, 52 students were absent. This was 13% of the students enrolled. How many students are enrolled in the school?

16) A family has a monthly income of $2,500. If 20% of the income is spent on food, how much is spent on food?

17) Alice answered 90% of the 60 problems on a test correctly. How many problems were answered correctly?

18) A road is to be 240 miles long. So far, 180 miles are completed. What percent of the road is completed?

19) A school has an enrollment of 600 students. If 12% of the students are absent, how many are present?

20) There are 40 7th graders in a club. If this is 80% of the club, then how many students are in the club?

1) _____

2) _____

3) _____

4) _____

5) _____

6) _____

7) _____

8) _____

9) _____

10) _____

11) _____

12) _____

13) _____

14) _____

15) _____

16) _____

17) _____

18) _____

19) _____

20) _____

Score: _____

Final Review – Pre-Algebra and Algebra

For 1 - 3, use the following sets to find the answers.

$$A = \{1,2,3,4,5\}, \quad B = \{2,3,4,6,8\}, \quad C = \{0,1,2,4,5,9\}$$

1. Find $A \cap B$ 2. Find $B \cup C$ 3. Find $A \cap C$

4. $-9 + 12 =$ 5. $-16 - 7 =$ 6. $-12 \times -3 =$

7. $-24 \div -3 =$ 8. $.21 + -.76 =$ 9. $-\dfrac{2}{5} + -\dfrac{1}{2} =$

10. $5^3 =$ 11. $\sqrt{49}$ 12. $3^3 + \sqrt{36} =$

13. $6 + 7 \times 3 - 5 =$ 14. $3^2(3 + 4) + 5 =$

15. $\dfrac{4^2 + 12}{5 + 3(2+1)} =$ 16. $2[(5 + 7) \div 3 + 6] =$

17. What property is illustrated below?

    $5 + 6 = 6 + 5$

18. What property is illustrated below?

    $7(6 + 5) = 7(6) + 7(5)$

19. Write 1,280,000,000 in scientific notation.

20. Write .00000653 in scientific notation.

1.
2.
3.
4.
5.
6.
7.
8.
9.
10.
11.
12.
13.
14.
15.
16.
17.
18.
19.
20.

Final Review – Pre-Algebra and Algebra

21. Write 6.09×10^7 as a conventional number.

22. Write $7.62 \times 10\text{-}6$ as a conventional number.

23. Write 18 to 10 as a fraction reduced to lowest terms.

24. Is $\frac{7}{8} = \frac{3}{4}$ a proportion? Why?

25. Solve the proportion.
$$\frac{7}{n} = \frac{3}{9}$$

26. The ratio of red marbles to blue marbles is five to two. If there are 15 red marbles, how many blue marbles are there?

27. Find 5% of 80.

28. Six is what percent of 24?

29. $8 = 25\%$ of what?

30. On a test with 30 questions, a student got 80% correct. How many questions did he get correct?

31. There are 35 fish in an aquarium. If 14 of them are goldfish, what percent of them are goldfish?

32. Six students get A's on a test. This is 20% of the class. How many are there in the class?

33. Find all the factors of 40.

34. Find the GCF of 60 and 40.

35. Find the first seven multiples of six.

36. Find the LCM of 8 and 12.

Use the number line to state the coordinates of the given points.

```
   A   D   I   N   J   P   E   T   B   S   K   R   F   L   Q   C   M   H   G
◄──┼───┼───┼───┼───┼───┼───┼───┼───┼───┼───┼───┼───┼───┼───┼───┼───┼───┼───┼──►
  -8  -7  -6  -5  -4  -3  -2  -1   0   1   2   3   4   5   6   7   8   9  10
```

37. A, E, B

38. N, H, T, D

39. I, F, P

40. H, M, S

21.	
22.	
23.	
24.	
25.	
26.	
27.	
28.	
29.	
30.	
31.	
32.	
33.	
34.	
35.	
36.	
37.	
38.	
39.	
40.	

Final Review – Pre-Algebra and Algebra

For 41 - 50, use the coordinate system to answer each question.

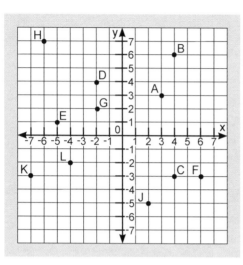

For 41-45, find the ordered pair associated with each point.

| 41. | F | 42. | E |
| 43. | C | 44. | L |

45.   B

For 46-52, find the point associated with each ordered pair.

46. (-6, 7)   47. (-2, 2)   48. (-4, -2)   49. (2, -5)   50. (-7, -3)

51. Find the slope of the line that passes through the points (1, 3) and (4, 5).

52. Find the slope of the line that passes through the points (-2, 5) and (6, 8).

For 53 through 54, make a table of 4 solutions and graph the points.
Connect them with a line.

53.

$y = x + 4$

x	y

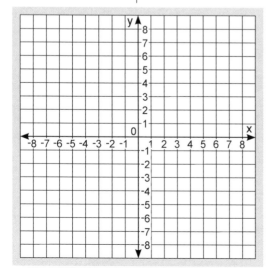

54.

$y = 2x + 2$

x	y

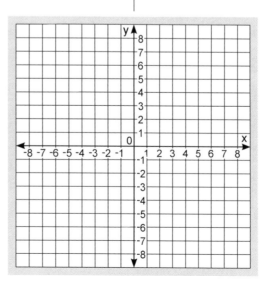

41.	
42.	
43.	
44.	
45.	
46.	
47.	
48.	
49.	
50.	
51.	
52.	
53.	
54.	

Final Review – Pre-Algebra and Algebra

Solve each equation and word problem.

55. $x + 3 = 12$

56. $3n = -45$

57. $\dfrac{n}{6} = 3$

58. $-5n = 15$

59. $2x + 3 = 15$

60. $5x - 2 = -17$

61. $\dfrac{n}{3} + -4 = 4$

62. $3(x + 4) = 24$

63. $3(x + 4) = -6$

64. $2x + 12 = 4x + 10$

65. Two more than three times a number is 29. Find the number.

66. Twice a number, less seven, is 17. Find the number.

67. A number divided by five, less six, is four. Find the number.

68. Sue has three times as much money as Jane. Together they have 64 dollars. How much does each have?

69. Al is seven years older than Maria. The sum of their ages is 51. What is each of their ages?

70. Six times Glen's age plus two equals four times his age plus 20. Find his age.

55.
56.
57.
58.
59.
60.
61.
62.
63.
64.
65.
66.
67.
68.
69.
70.

Final Review – Pre-Algebra and Algebra

Use the following information to answer 71 - 74.

There are 6 green marbles, 5 red marbles, 4 white marbles, and 1 blue marble in a can. What is the probability for each of the following?

71. a red marble   72. a green marble

73. a green or blue marble   74. not a red marble

Use the spinner to find the probability of spinning
once and landing on each of the following.

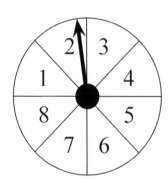

75. a seven.

76. an even number.

77. a number greater than four.

78. a one, a three, or a five.

Arrange the data in order from least to greatest, then answer the questions.

2, 7, 3, 10, 3

79. What is the range?   80. What is the mode?

81. What is the mean?   82. What is the median?

4, 10, 4, 2, 6, 14, 16

83. What is the mode?   84. What is the mean?

85. What is the range?   86. What is the median?

71.
72.
73.
74.
75.
76.
77.
78.
79.
80.
81.
82.
83.
84.
85.
86.

Final Review – Pre-Algebra and Algebra

Solve each of the following problems.

87. $\dfrac{7}{8}$

$+ \dfrac{3}{8}$

88. $7\dfrac{1}{4}$

$- 3\dfrac{3}{4}$

89. $3\dfrac{3}{5}$

$+ 2\dfrac{1}{10}$

90. $\dfrac{5}{8} \times 3\dfrac{1}{5} =$

91. $2\dfrac{1}{2} \times 3\dfrac{1}{2} =$

92. $2\dfrac{1}{3} \div \dfrac{1}{2} =$

93. $5\dfrac{1}{2} \div 1\dfrac{1}{2} =$

94. $.6 + 7.62 + 5.2 + 6 =$

95. $6.3 - 1.275 =$

96. $72 - 1.68 =$

97. 2.19

$\times\ 7$

98. $.36$

$\times\ 1.2$

99. $5\overline{)6.7}$

100. $.15\overline{).0045}$

87.
88.
89.
90.
91.
92.
93.
94.
95.
96.
97.
98.
99.
100.

Page 11

1. 437
2. 735
3. 216
4. 12

S. 917
S. 1,484
1. 111
2. 1,102
3. 1,998
4. 1,186
5. 1,997
6. 675
7. 1,115
8. 207
9. 1,693
10. 105
Problem: 100 students

Page 12

1. 95
2. 823
3. 1,844
4. 95

S. 6,952
S. 29,994
1. 3,470,346
2. 51,422
3. 155,804
4. 1,879,958
5. 42,452
6. 83,749
7. 14,935
8. 31,116
9. 17,276
10. 82,646
Problem: 4,540,028

Page 13

1. 10,321
2. 40,719
3. 98,477
4. 94,257

S. 481
S. 2,461
1. 241
2. 399
3. 6,385
4. 5,877
5. 647
6. 3,193
7. 38
8. 3,070
9. 6,399
10. 339
Problem: 245 students

Page 14

1. 1,166
2. 574
3. 10,511
4. 5,649

S. 434
S. 221
1. 16
2. 435
3. 3,608
4. 264
5. 6,366
6. 1,468
7. 5,232
8. 62,420
9. 12,868
10. 1,999
Problem: $27,044

Page 15

1. 64
2. 6,429
3. 1,152,079
4. 4,533

S. 29,710
S. 3,655
1. 547
2. 573
3. 10,103
4. 7,243,008
5. 5,665
6. 1,196
7. 3,418
8. 1,845
9. 616
10. 90,558
Problem: 212 seats

Page 16

1. 4,385
2. 999
3. 572
4. 5,713

S. 1,269
S. 14,070
1. 201
2. 444
3. 2,292
4. 18,852
5. 16,288
6. 20,562
7. 56,406
8. 53,501
9. 58,476
10. 123,372
Problem: 2,190 days

Page 17

1. 2,052
2. 14,238
3. 5,439
4. 37,053

S. 1,058
S. 6,132
1. 4,526
2. 1,200
3. 1,692
4. 13,350
5. 7,110
6. 29,052
7. 448
8. 988
9. 67,528
10. 15,680
Problem: 832 desks

Page 18

1. 1,058
2. 14,472
3. 5,568
4. 997

S. 30,888
S. 337,666
1. 50,095
2. 167,564
3. 29,029
4. 99,138
5. 199,584
6. 78,900
7. 152,703
8. 130,662
9. 432,600
10. 155,644
Problem: 35,260 cars

Page 19

1. 2,142
2. 864
3. 10,844
4. 1,538

S. 7,584
S. 255,492
1. 162
2. 4,221
3. 32,508
4. 4,536
5. 14,202
6. 155,606
7. 5,676
8. 448,800
9. 9,020
10. 214,624

Problem: 12,000 sheets

Page 20

1. 249
2. 1,920
3. 138,863
4. 962

S. 5 r1
S. 9 r6
1. 8 r2
2. 5 r3
3. 12 r3
4. 15 r3
5. 12 r1
6. 7 r1
7. 9 r3
8. 24 r1
9. 12 r1
10. 18 r1

Problem: 12 boxes

Page 21

1. 6 r3
2. 15 r4
3. 14,938
4. 5,232

S. 144
S. 39 r3
1. 317 r1
2. 409 r1
3. 132
4. 136 r1
5. 145 r2
6. 158 r2
7. 121
8. 98 r4
9. 122 r3
10. 158 r5

Problem: 36 seats, 1 leftover

Page 22

1. 26 r1
2. 92 r5
3. 1,058
4. 152

S. 2,354
S. 863
1. 566
2. 3,371 r1
3. 1,122
4. 386 r3
5. 1,311
6. 1,498 r3
7. 10,453 r1
8. 11,980
9. 10,192 r1
10. 5,351

Problem: 144 boxes

Page 23

1. 203 r1
2. 7,416
3. 543
4. 22,059

S. 81 r2
S. 1,071
1. 50 r2
2. 1,500
3. 2,104 r2
4. 418 r5
5. 223 r2
6. 89 r4
7. 302 r5
8. 798 r5
9. 1,001 r1
10. 400 r7

Problem: 106 tickets

Page 24

1. 116 r6
2. 1,238
3. 271,800
4. 1,000 r3

S. 6 r7
S. 133 r22
1. 4 r62
2. 6 r39
3. 9 r13
4. 14 r27
5. 58 r23
6. 71 r3
7. 166 r24
8. 103 r11
9. 95 r1
10. 50 r26

Problem: 86 boxes

Page 25

1. 1,211
2. 156
3. 21,600
4. 1,167 r3

S. 21 r21
S. 31 r2
1. 22 r25
2. 21 r15
3. 10 r63
4. 7 r41
5. 21 r10
6. 31 r23
7. 11 r42
8. 11 r28
9. 45 r1
10. 21 r5

Problem: 13 classes

Page 26

1. 358
2. 17 r14
3. 21 r20
4. 20 r11

S. 3 r71
S. 19 r4
1. 5 r7
2. 6
3. 55 r7
4. 8 r10
5. 5 r3
6. 63 r4
7. 19 r3
8. 19 r31
9. 20 r1
10. 43 r13

Problem: 864 eggs

Page 27

1. 3 r6
2. 13 r22
3. 9 r36
4. 6 r19

S. 212 r22
S. 207 r15
1. 115
2. 234 r3
3. 102
4. 133 r10
5. 303 r10
6. 306
7. 73 r4
8. 40 r6
9. 45 r17
10. 41 r24

Problem: 125 pounds

Page 28

1. 24
2. 298 r4
3. 135 r36
4. 42 r17

S. 96 r40
S. 59 r9
1. 43 r1
2. 276 r4
3. 1,025 r2
4. 18 r7
5. 2 r56
6. 84 r19
7. 255 r9
8. 114 r11
9. 36 r26
10. 220 r9

Problem: 16 gallons

Page 29 Review

1. 1,011
2. 2,919
3. 80,653
4. 10,433
5. 8,447
6. 376
7. 3,915
8. 4,415
9. 2,322
10. 6,323
11. 228
12. 30,612
13. 2,438
14. 22.572
15. 232,858
16. 141 r2
17. 282 r5
18. 25 r19
19. 214 r4
20. 56 r9

Page 30

1. 434
2. 1,156
3. 1,387 r1
4. 1,296

S. 1/4
S. 5/6
1. 3/4, 6/8
2. 1/3
3. 2/4, 1/2
4. 3/8
5. 5/8
6. 2/3
7. 5/6
8. 1/8
9. 2/6, 1/3
10. 7/8

Problem: 384 crayons

Page 31

1. 2,142
2. 1,114
3. 494
4. 15 r17

S. 1/2
S. 3/4
1. 4/5
2. 3/4
3. 1/2
4. 4/5
5. 2/3
6. 2/3
7. 3/5
8. 5/8
9. 5/6
10. 3/4

Problem: 60 crayons

Page 32

1. 3/8
2. 2/3
3. 21 r7
4. 65,664

S. 1 ¾
S. 1 ½
1. 2 ½
2. 1 3/7
3. 2
4. 4 4/5
5. 4 1/2
6. 3
7. 4 1/2
8. 6 1/3
9. 1 1/2
10. 2 2/5

Problem: 2,828 more

Page 33

1. 1 2/7
2. 1 1/4
3. 5/7
4. 3/4

S. 3/5
S. 1 1/7
1. 3/4
2. 3/4
3. 3/5
4. 1/2
5. 1 1/4
6. 1 1/2
7. 1 2/3
8. 1 3/4
9. 1 3/4
10. 1 2/3

Problem: 2/3

Page 34

1. 3/5
2. 12 r9
3. 1 1/5
4. 368

S. 5 1/2
S. 6
1. 8
2. 7 1/2
3. 8
4. 8 1/6
5. 12 1/3
6. 6 1/5
7. 9 1/5
8. 12 1/2
9. 10
10. 8 1/7
Problem: 2 1/8 cups

Page 35

1. 5/6
2. 3 4/5
3. 1 1/2
4. 7 1/5

S. 1/2
S. 1/2
1. 1/2
2. 4/5
3. 3/4
4. 2/3
5. 3/4
6. 4/11
7. 1/2
8. 5/7
9. 5/12
10. 1/4
Problem: 3/5 miles

Page 36

1. 1/2
2. 1 1/5
3. 5 1/2
4. 943

S. 2 1/2
S. 2 2/3
1. 6 1/4
2. 4 1/2
3. 4 11/15
4. 6 2/3
5. 4 2/5
6. 3 2/5
7. 2
8. 2 3/5
9. 3 3/5
10. 2 3/5
Problem: 6 1/3 hours

Page 37

1. 1/2
2. 1 1/4
3. 8 1/3
4. 12 1/2

S. 3 2/5
S. 6 ¼
1. 3 3/7
2. 3 2/5
3. 6 1/3
4. 3 1/10
5. 4 7/8
6. 3 5/8
7. 3 2/9
8. 1/2
9. 3 7/10
10. 4 2/5
Problem: 2 1/8 yards

Page 38

1. 5/8
2. 1
3. 1 2/5
4. 4/5

S. 8 1/5
S. 4 2/3
1. 1 1/9
2. 3/4
3. 11 1/5
4. 4 2/3
5. 5 1/2
6. 4 2/3
7. 8 1/2
8. 4 2/3
9. 3 2/5
10. 2
Problem: 8 2/3 pounds

Page 39

1. 1 2/5
2. 1/2
3. 4 1/4
4. 3 3/5

S. 12
S. 24
1. 15
2. 18
3. 14
4. 24
5. 36
6. 20
7. 39
8. 60
9. 48
10. 36
Problem: 750 mph

Page 40

1. 200 r2
2. 1,548
3. 352
4. 461

S. 7/12
S. 1/2
1. 5/9
2. 1/6
3. 1 1/6
4. 11/15
5. 5/12
6. 1 1/14
7. 1 1/2
8. 17/22
9. 1/14
10. 1 5/36
Problem: 3 5/8 gallons

Page 41

1. 188 r2
2. 3/4
3. 1 1/10
4. 5/12

S. 7 11/12
S. 8 1/10
1. 8 1/6
2. 10 3/4
3. 8 1/8
4. 8 1/2
5. 15 3/4
6. 9 13/ 20
7. 5 3/10
8. 11 1/18
9. 8 8/15
10. 11 11/12
Problem: 4,224 parts

Page 42

1. 13/14
2. 10
3. 4 2/5
4. 1 2/3

S. 3 1/20
S. 2 5/6
1. 2 5/8
2. 7 1/2
3. 2 7/12
4. 4 5/8
5. 13/14
6. 5 13/16
7. 3 2/9
8. 3 5/6
9. 4 13/20
10. 2 3/8
Problem: 24 students

Page 43

1. 4/5
2. 7 1/4
3. 12
4. 1 1/30

S. 1 3/7
S. 10 ¼
1. 7/18
2. 13/18
3. 4 2/5
4. 8 5/8
5. 11 1/4
6. 4 3/4
7. 8 7/15
8. 3 1/2
9. 11/16
10. 8 7/12
Problem: 3 3/4

Page 44

1. 121
2. 26,100
3. 221
4. 6,268

S. 15/28
S. 12/25
1. 2/63
2. 1/5
3. 2 1/10
4. 14/27
5. 2/3
6. 1 1/15
7. 1 1/5
8. 6/35
9. 2/5
10. 1 1/14
Problem: 13/20 pounds

Page 45

1. 4/7
2. 1 13/15
3. 3/4
4. 1 1/3

S. 3/7
S. 1 1/2
1. 3/8
2. 1/10
3. 7/18
4. 3 1/3
5. 3/10
6. 2/3
7. 2/5
8. 9/20
9. 1 1/7
10. 10/21
Problem: 390 seats

Page 46

1. 3/7
2. 14/27
3. 13/15
4. 1/12

S. 9
S. 3 1/3
1. 6
2. 4
3. 20
4. 1 1/7
5. 13 1/2
6. 2 1/2
7. 3 1/2
8. 7 1/2
9. 15
10. 5 3/5
Problem: 24 girls

Page 47

1. 12 r40
2. 12
3. 6 2/3
4. 7/2

S. 3/4
S. 3
1. 1 7/8
2. 3 1/9
3. 16
4. 3
5. 6
6. 3 3/8
7. 8 1/8
8. 15
9. 11 2/3
10. 1 3/10
Problem: 14 miles

Page 48

1. 12/35
2. 27/100
3. 18
4. 8 2/3

S. 3/10
S. 7 1/2
1. 7/10
2. 1/6
3. 21
4. 5 1/7
5. 1 7/8
6. 1 5/6
7. 17
8. 2 1/3
9. 15 1/5
10. 8 1/8
Problem: 22 1/2 tons

Page 49

1. 14/15
2. 1/4
3. 7
4. 1 7/9

S. 1 1/3
S. 3/7
1. 1/6
2. 1 1/7
3. 4/13
4. 1/13
5. 2 1/2
6. 7
7. 1/9
8. 4 1/2
9. 2/9
10. 1/15
Problem: 45 dollars

Page 50

1. 1/9
2. 3 1/2
3. 3/11
4. 3 1/5

S. 1 1/7
S. 1 3/4
1. 2 1/4
2. 1 1/2
3. 7 1/2
4. 9
5. 4 2/3
6. 9
7. 1/2
8. 2 3/4
9. 3
10. 1 7/15
Problem: 7 pieces

Page 51 Review

1. 4/5
2. 1 1/3
3. 13/15
4. 8 2/9
5. 10 1/8
6. 1/2
7. 4 4/5
8. 4 2/5
9. 6 1/4
10. 5 14/15
11. 8/21
12. 3/26
13. 27
14. 1 7/8

15. 8 1/6
16. 1 1/2
17. 7
18. 2 4/9
19. 3 1/3
20. 2 4/7

Page 52

1. 1,876
2. 1 4/15
3. 428
4. 7/12

S. Two and six tenths
S. Thirteen and sixteen thousandths
1. seventy-three hundredths
2. four and two thousandths
3. one hundred thirty-two and six tenths
4. one hundred thirty-two and six hundredths
5. seventy-two and six thousand three hundred ninety-five ten thousandths
6. seventy-seven thousandths
7. nine and eighty-nine hundredths
8. six and three thousandths
9. seventy-two hundredths
10. one and six hundred sixty-six thousandths
Problem: 21 boys

Page 53

1. 8,424
2. 6/11
3. 5
4. 1 2/3

S. 6.04
S. 306.15
1. 9.8
2. 46.013
3. .0326
4. 50.039
5. .00008
6. .000004
7. 12.0036
8. 16.024
9. 23.5
10. 2.017
Problem: 3 2/5 degrees

Page 54

1. 1 1/2
2. 3/4
3. 2
4. 8 1/3

S. 7.7
S. 9.007
1. 16.32
2. .0097
3. 72.09
4. 134.0092
5. .016
6. 44.00432
7. 3.096
8. 4.901
9. 3.000901
10. .01763
Problem: 36 inches

Page 55

1. 13/15
2. 5 2/3
3. 1
4. 4/33

S. 1 43/100
S. 7 6/1,000
1. 173 16/1,000
2. 16/100,000
3. 7 14/1,000,000
4. 19 936/1,000
5. 9163/100,000
6. 77 8/10
7. 13 19/1,000
8. 72 9/10,000
9. 99/100,000
10. 63 143/1,000,000

Problem: 120 seats

Page 56

1. 5/6
2. 4 3/8
3. One and nineteen thousandths
4. 72 8/1,000

S. <
S. >
1. <
2. >
3. >
4. >
5. >
6. <
7. <
8. >
9. <
10. <

Problem: 8 miles

Page 57

1. 1 3/4
2. 1 1/4
3. 1
4. 14 2/5

S. 18.82
S. 8.84
1. 41.353
2. 14.431
3. 29.673
4. 1.7
5. 19.26
6. 146.983
7. 1.7
8. 25.044
9. 42.055
10. 31.09

Problem: 15.8 inches

Page 58

1. 13.726
2. 15.76
3. 1 3/4
4. 72.009

S. 13.84
S. 7.48
1. 5.894
2. 2.293
3. 1.16
4. 11.13
5. .053
6. 3.9577
7. 2.373
8. 3.684
9. 8.628
10. 26.765

Problem: 1.3 seconds

Page 59

1. 2/3
2. 3/4
3. 1 1/4
4. 2 1/5

S. 18.38
S. 3.687
1. 23.714
2. 6.17
3. 16.57
4. 8.24
5. 2.08
6. 19.56
7. 2.2
8. 4.487
9. 10.396
10. 25.7

Problem: $48.72

Page 60

1. 216
2. 1,472
3. 4,807
4. 265,608

S. 7.38
S. 36.8
1. 1.929
2. 14.64
3. 6.88
4. 5.664
5. 22.4
6. 55.2
7. 328.09
8. 10.024
9. 29.93
10. 1.242

Problem: 18.9 miles

Page 61

1. 217.2
2. 4.32
3. 1 1/8
4. 1 1/16

S. 2.52
S. 7.776
1. 11.52
2. .4598
3. .21186
4. .874
5. .1421
6. 9.7904
7. .0024
8. 1.904
9. 149.225
10. .0738

Problem: 11.25 tons

Page 62

1. 3 2/3
2. 1 1/2
3. 1 1/3
4. 8

S. 32
S. 7,390
1. 93.6
2. 72,600
3. 160
4. 736.2
5. 7,280
6. 70
7. 37.6
8. 390
9. 73.3
10. 76.3
Problem: $9,500.00

Page 63

1. 7 1/2
2. 7
3. 14/17
4. 10

S. 2.394
S. 15.228
1. 3.22
2. .464
3. .48508
4. 26
5. 3.156
6. 2,630
7. .0108
8. 3.575
9. 55.11
10. .04221
Problem: $41.37

Page 64

1. 19
2. 202
3. 1,001
4. 111

S. .44
S. 1.8
1. 19.7
2. 3.21
3. .57
4. 3.7
5. 6.08
6. 40.9
7. .324
8. .16
9. 2.12
10. 2.04
Problem: 21 pieces

Page 65

1. 2.18
2. .3
3. 18.83
4. 4.833

S. .0027
S. .019
1. .0007
2. .012
3. .056
4. .036
5. .043
6. .063
7. .023
8. .022
9. .003
10. .068
Problem: $5.51

Page 66

1. 34.2
2. .0932
3. 3
4. 3

S. .34
S. .06
1. .065
2. .62
3. 2.05
4. .15
5. .04
6. .04
7. .12
8. 1.575
9. .06
10. .418
Problem: 28

Page 67

1. .075
2. .026
3. 930
4. 9,300

S. 3.9
S. .24
1. 8
2. 170
3. .42
4. 80
5. 5.4
6. 3.2
7. 3.7
8. 3.2
9. 57.5
10. 9.2
Problem: 55.25 mph

Page 68

1. .8
2. 50
3. .38
4. 4

S. .5
S. .625
1. .6
2. .25
3. .4
4. .875
5. .55
6. .52
7. .625
8. .2
9. .2
10. .7
Problem: $15.00

Page 69

1. 10.96
2. 1.57
3. 2.184
4. .875

S. .005
S. 40
1. .76
2. .074
3. .025
4. 4.5
5. .24
6. 284
7. 33.1
8. 16.2
9. .4
10. .625
Problem: 117.25 pounds

Solutions

Page 70 Review

1. 12.283
2. 10.36
3. 29.1
4. 20.6
5. 4.73
6. 4.57
7. 9.36
8. 54.4
9. .752
10. 1.39956
11. 236
12. 2,700
13. 1.34
14. 1.46
15. .65
16. 400
17. .05
18. .013
19. .875
20. .44

Page 71

1. 1 1/2
2. 3
3. 1 1/6
4. 7/15

S. 17%
S. 90%
1. 6%
2. 99%
3. 30%
4. 64%
5. 67%
6. 1%
7. 70%
8. 14%
9. 80%
10. 62%
Problem: 90 pounds

Page 72

1. 7%
2. 90%
3. 1 1/4
4. 7.75

S. 37%
S. 70%
1. 93%
2. 2%
3. 20%
4. 9%
5. 60%
6. 66%
7. 89%
8. 60%
9. 33%
10. 80%
Problem: 12.8 fluid ounces

Page 73

1. 3/4
2. 1 1/8
3. 3 7/10
4. 6 1/10

S. .2, 1/5
S. .09, 9/100
1. .16, 4/25
2. .06, 3/50
3. .75, 3/4
4. .4, 2/5
5. .01, 1/100
6. .45, 9/20
7. .12, 3/25
8. .05, 1/20
9. .5, 1/2
10. .13, 13/100
Problem: 3/4

Page 74

1. 1.8
2. .8
3. 1.872
4. 9.62

S. 17.5
S. 150
1. 4.32
2. 51
3. 15
4. 112.5
5. 32
6. 80
7. 10
8. 216
9. 112.5
10. 13.2
Problem: 1 1/4 gallons

Page 75

1. 11.05
2. .8
3. .03
4. 12

S. 3
S. 30
1. $56
2. $1,800
3. 9
4. $750
5. 16
6. $16,000
7. $4,200
8. $3.50/$53.50
9. 138
10. $1,650
Problem: 208.75 miles

Page 76

1. 88 r4
2. 900
3. 3,918
4. 96

S. 20%
S. 80%
1. 60%
2. 50%
3. 10%
4. 75%
5. 75%
6. 60%
7. 25%
8. 80%
9. 75%
10. 20%
Problem: 64 people

Page 77

1. 12.57
2. 8.857
3. 44.814
4. 4.7

S. 25%
S. 75%
1. 25%
2. 80%
3. 50%
4. 90%
5. 60%
6. 75%
7. 75%
8. 75%
9. 80%
10. 95%
Problem: $235

Page 78

1. 26.4
2. 60%
3. 27
4. 34

S. 75%
S. 40%
1. 80%
2. 75%
3. 25%
4. 80%
5. 95%
6. 20%
7. 48%
8. 20%
9. 98%
10. 75%
Problem: 24 correct

Page 79

1. 7%
2. 90%
3. 30%
4. .24, 6/25

S. 9
S. 25%
1. 3.6
2. 57.6
3. 75%
4. 80%
5. 77.5
6. 6
7. 25%
8. 10%
9. 450
10. 51.92
Problem: 80%

Page 80

1. 1/6
2. 5 3/4
3. 3 4/15
4. 2 2/5

S. 32
S. 25%
1. 300
2. 60%
3. $14.40
4. 80%
5. 75%
6. 14.4
7. 240
8. 10%
9. $1,200
10. 20%, 80%
Problem: 84

Page 81 Review

1. 17%
2. 3%
3. 70%
4. 19%
5. 60%
6. .09, 9/100
7. .14, 7/50
8. .8, 4/5
9. 12.8
10. 138

11. 72
12. 60%
13. 80%
14. 25%
15. 20%
16. 25%
17. $80
18. 180
19. 45%
20. 75%

Page 82

1. 8.22
2. 13.14
3. .0492
4. .6

S. answers vary
S. answers vary
1. answers vary
2. answers vary
3. F, D, E points
4. $\overleftrightarrow{AC}$, $\overleftrightarrow{BC}$
5. $\overleftrightarrow{FE}$, $\overleftrightarrow{FD}$
6. $\overrightarrow{AB}$
7. $\overline{FD}$, $\overline{ED}$, $\overline{FE}$
8. answers vary
9. answers vary
10. point E
Problem: 25 hours

Page 83

1. 1 2/5
2. 1/2
3. 1 1/2
4. 4

S. answers vary
S. answers vary
1. answers vary
2. answers vary
3. answers vary
4. answers vary
5. answers vary
6. answers vary
7. answers vary
8. answers vary
9. answers vary
10. ∠IHJ
Problem: 5 pieces

Page 84

1. 75%
2. 33.75
3. 45
4. 70%

S. answers vary
S. answers vary
1. answers vary
2. answers vary
3. acute
4. obtuse
5. right
6. straight
7. answers vary
8. answers vary
9. answers vary
10. answers vary
Problem: .34 meters

Page 85

1. 1 2/7
2. 23/3
3. 3/5
4. 8 3/4

S. acute, 20°
S. obtuse, 110°
1. right, 90°
2. obtuse, 160°
3. acute, 20°
4. acute, 70°
5. obtuse, 130°
6. acute, 50°
7. straight, 180°
8. obtuse, 160°
9. right, 90°
10. obtuse, 130°
Problem: 208 students

Page 86

1. 75%
2. 90%
3. 2.6
4. 40%

S. acute
S. acute
1. acute
2. acute
3. obtuse
4. obtuse
5. acute
6. acute
7. right
8. obtuse
9. obtuse
10. acute
Problem: 266 seats

Page 87

1. DEF, acute
2. FGH, obtuse
3. JKL, right
4. parallel

S. rectangle/parallelogram
S. triangle
1. square; rectangle; parallelogram
2. rectangle; parallelogram
3. trapezoid
4. triangle
5. trapezoid
6. square; rectangle and parallelogram
7. parallelogram
8. rectangle; parallelogram
9. triangle
10. trapezoid
Problem: $30.00

Page 88

1. 233 r23
2. 1,728
3. 1,155
4. 6,812

S. scalene/right
S. equilateral/acute
1. scalene; obtuse
2. isosceles; acute
3. isosceles; right
4. scalene; acute
5. equilateral; acute
6. scalene; obtuse
7. scalene; right
8. isosceles; acute
9. equilateral; acute
10. isosceles; right
Problem: 3 buses

Page 89

1. scalene
2. right
3. isosceles/acute
4. 45

S. 34 ft.
S. 30 ft.
1. 47 ft.
2. 48 ft.
3. 33 ft.
4. 54 ft.
5. 70 ft.
6. 41 ft.
7. 225 mi.
8. 34 ft.
9. 86 ft.
10. 34 ft.
Problem: 190 ft.

Page 90

1. 46 ft.
2. 56 ft.
3. 20%
4. 5/12

S. diameter
S. answers vary
1. radius
2. chord
3. $\overline{CD}$; $\overline{DE}$; $\overline{DG}$; $\overline{DF}$
4. $\overline{AB}$; $\overline{GF}$; $\overline{CE}$
5. 8 feet
6. point P
7. $\overline{RY}$; $\overline{VT}$; $\overline{XS}$
8. 48 feet
9. $\overline{PX}$; $\overline{PS}$; $\overline{PZ}$
10. $\overline{XS}$

Problem: 16 miles

Page 91

1. 350
2. 12.81
3. 4.1
4. 24.04

S. 12.56 ft.
S. 31.4 ft.
1. 18.84 ft.
2. 25.12 ft.
3. 28.26 ft.
4. 87.92 ft. or 88 ft.
5. 37.68 ft.
6. 31.4 ft.
7. 12.56 ft.

Problem: 75.36 ft.

Page 92

1. 19.52
2. 1
3. 3
4. 227.5

S. 169 sq. ft.
S. 165 sq. ft.
1. 84 sq. ft.
2. 400 sq. ft.
3. 132 sq. ft.
4. 10.75 sq. ft.
5. 6 sq. ft.
6. 625 sq. ft.
7. 6 1/4 sq. ft.

Problem: 182 sq. ft.

Page 93

1. 224 sq. ft.
2. 256 sq. ft.
3. 25.12 ft.
4. 43.96 ft.

S. 78 sq. ft.
S. 77 sq. ft.
1. 54 sq. ft.
2. 176 sq. ft.
3. 17 1/2 sq. ft.
4. 91 sq. ft.
5. 84 sq. ft.
6. 117 sq ft.
7. 71 1/2 sq. ft.

Problem: 94.2 ft.

Page 94

1. 800 sq. ft.
2. 45 sq. ft.
3. 98 sq. ft.
4. 169 sq. ft.

S. 50.24 sq. ft.
S. 113.04 sq. ft.
1. 78.5 sq. ft.
2. 615.44 sq. ft. or 616 sq. ft.
3. 12.56 sq. ft.
4. 200.96 sq. ft.
5. 78.5 sq. ft.
6. 113.04 sq. ft.
7. 153.86 sq. ft. or 154 sq. ft.

Problem: 75.36 ft.

Page 95

1. 52 ft.
2. 168 sq. ft.
3. 18.84 ft.
4. 28.26 sq. ft.

S. P=38 ft.   A= 84 sq.ft.
S. P=23 ft.   A= 20 sq. ft.
1. P= 48 ft.   A= 144 sq. ft.
2. P= 44 ft.   A= 120 sq. ft.
3. P= 38 ft.   A= 72 sq. ft.
4. C= 18.84 ft.   A= 28.26 sq. ft.
5. C= 43.96 ft. (44 ft.)   A= 153.86 sq. ft. (154 sq. ft.)
6. P= 24 ft.   A= 24 sq. ft.
7. P= 32 ft.   A= 64 sq. ft.

Problem: 216 sq. ft.

Page 96

1. 1/10
2. 1 1/6
3. 10 1/2
4. 5

S. rectangular prism
6 faces; 12 edges; 8 vertices

S. square pyramid
5 faces; 8 edges; 5 vertices

1. cone, 1, 1, 1
2. cube, 6, 12, 8
3. cylinder, 2, 2, 0
4. triangular pyramid, 4, 6, 4
5. triangular prism, 5, 9, 6
6. sphere
7. 1

Problem: 40% girls; 60% boys

Page 97 Review

1. answers vary
2. answers vary
3. answers vary
4. answers vary
5. answers vary
6. answers vary
7. answers vary
8. answers vary
9. sides, equilateral,
   angles, acute
10. sides, scalene,
    angles, right
11. 28 feet
12. 28.26 feet
13. 153.86 sq. ft. (154 sq. ft.)
14. 98 sq. ft.
15. 126 sq. ft.
16. 58 1/2 sq. ft.
17. 256 sq. ft.
18. square, pyramid, 5, 8, 5
19. 140 sq. ft.
20. 384 feet

Page 98

1. 78 sq. ft
2. 40.82 ft.
3. 75%
4. 12.75

S. 3
S. -21
1. 14
2. -18
3. -14
4. -31
5. 24
6. 37
7. -168
8. -13
9. -34
10. -9
Problem: 336 students

Page 99

1. 2
2. -2
3. -34
4. 28.26 sq. ft.

S. -4
S. -7
1. -2
2. -7
3. -8
4. 14
5. -2
6. 10
7. -22
8. -13
9. -25
10. -84
Problem: won 75%, lost 25%

Page 100

1. 84 ins.
2. 1
3. 9
4. -3

S. -14
S. -15
1. 12
2. -3
3. 9
4. -28
5. 46
6. 21
7. -10
8. -11
9. -103
10. 15
Problem: $9.03

Page 101

1. -15
2. 3
3. 34
4. -108

S. -40
S. 15
1. -53
2. 11
3. -15
4. 15
5. 12
6. 3
7. -33
8. 15
9. 101
10. -681
Problem: 13 packages

Page 102

1. 3 3/4
2. 4
3. 12
4. 8

S. 48
S. -126
1. 68
2. -64
3. 288
4. -368
5. -736
6. -133
7. 21
8. 380
9. -256
10. 192
Problem: -10º

Page 103

1. 5
2. 10
3. 27
4. 252

S. 42
S. 54
1. -24
2. -48
3. 120
4. 360
5. -6
6. -24
7. -32
8. 72
9. 72
10. 330
Problem: $22.50

Page 104

1. 101 r1
2. 60
3. 86.1
4. 14.81

S. -3
S. 6
1. -16
2. 48
3. 15
4. -26
5. 22
6. -24
7. -6
8. -34
9. 13
10. -19
Problem: -26º

Page 105

1. -9
2. 8
3. -90
4. -24

S. 1
S. -12
1. -2
2. 3
3. -4
4. -36
5. -3
6. -2
7. 1
8. -1
9. 3
10. 2
Problem: 498 points

Page 106 Review

1. -2
2. 2
3. -16
4. -1
5. -19
6. -2
7. 13
8. -12
9. -27
10. -1
11. -48
12. 76
13. 56
14. 48
15. -9
16. 42
17. 16
18. 3
19. -2
20. 20

Solutions

Page 107

1. 153.86 sq. ft.
2. 9.75
3. 1 1/2
4. 1 3/4

S. April
S. 10
1. July
2. July
3. June
4. 7° -8°
5. April
6. March; April
7. May; July
8. 13°
9. 3° -4°
10. March, April, July
Problem: 37.68 yards

Page 108

1. 75%
2. 70%
3. 192 sq. ft.
4. 220 r3

S. Winston/Auberry
S. 1,050
1. 1,200
2. 300
3. 350
4. 3,300
5. 300
6. 450
7. 1,450
8. 300
9. Sun City
10. 1,250
Problem: 20 parts

Page 109

1. 39 sq. ft.
2. scalene/obtuse
3. 2,365
4. 43.38

S. 90
S. 10
1. Tests, 4, 6, 7
2. 20
3. approx. 91
4. 4
5. 80, 85
6. 90
7. 15
8. 15
9. 3
10. improved
Problem: $26.25

Page 110

1. 1 2/5
2. 3/4
3. 1 6/7
4. 1 2/3

S. July/September
S. 100
1. 700
2. 100
3. May
4. June
5. 100
6. July, Sept.
7. 1,100
8. 200
9. 100
10. April, May
Problem: 85

Page111

1. -44
2. 15
3. -21
4. 30

S. 23%
S. 70%
1. 33%
2. $460
3. $200
4. $400
5. 68%
6. 32%
7. $24,000
8. 80%
9. car, clothing
10. other expenses
Problem: 1,225 miles

Page 112

1. 3/4
2. 75%
3. 75%
4. -17

S. 1/6
S. 1/4
1. 5
2. 10
3. 8
4. 9/24 = 3/8
5. 16/24 = 2/3
6. 90
7. 6:00 A.M.
8. 2:30 P.M.
9. 40
10. 4/24 = 1/6
Problem: 168 sq.ft.

Page 113

1. 6,332
2. 30,628
3. 932
4. 1,003

S. 6,000
S. 4,500
1. 1989
2. 13,500
3. 1989; 1991
4. 7,000
5. 11,000
6. $150,000
7. $600,000
8. 5,000
9. 19,500
10. 3,000
Problem: $131

Page 114

1. 64 ins.
2. 18.84 ft.
3. 390 sq. ins.
4. 44 sq. ft.

S. 1990
S. 10 hours
1. 45
2. approx. 5
3. 2,000
4. 1960
5. 1950, 1960
6. approx. 2
7. 8
8. approx. 12
9. approx. 5 hours
10. approx, 10 hours
Problem: $26.25

Solutions

Page 115 Review

1. Grant Falls
2. 525-550 feet
3. 100 feet
4. Snake Falls
5. Morton Falls
6. 13%
7. 20%
8. $600
9. $300
10. 41%
11. 71° - 72°
12. 30°
13. June
14. August
15. approx. 10°
16. 60,000
17. 40,000
18. 200,000
19. 180,000 pounds
20. salmon, cod, snapper

Page 116

1. 1,211
2. 539
3. 11,584
4. 162 r3

S. 106
S. 2,485 miles
1. 6,976
2. $209
3. 1,718
4. 267 votes
5. 9,000 miles
6. 59 mph
7. 276 miles
8. 833 feet
9. 272 pieces
10. 5, 172 books

Page 117

1. 1 1/10
2. 5/8
3. 7 1/2
4. 5

S. 3 17/20 cups
S. 11 pieces
1. 8 1/4 pounds
2. $40
3. 30 miles
4. 29 1/4 minutes
5. 34 feet
6. 2 2/3 pounds
7. 1 1/2 pounds
8. 13 3/4 hours
9. 125 miles
10. 16 tires

Page 118

1. $13.69
2. $3.44
3. $23.52
4. $2.19

S. $1.44
S. 540 mph
1. $ .41
2. $429.70
3. 2.34 sq. miles
4. $1.14
5. 192 miles
6. 6 pounds
7. $10.55
8. 264.85 pounds
9. $.89
10. 8 gallons

Page 119

1. 36.8
2. 5
3. 153.86 sq. ft.
4. 144 sq. ft.

S. 4 pieces
S. 16 cans
1. 556 feet
2. $3.36
3. 89
4. 8.025 inches
5. $28
6. 6 1/4 feet
7. $36.36
8. 40 students
9. 24,197
10. 270 cows

Page 120

1. 36 r1
2. 234 r2
3. 63 r7
4. 210 r15

S. 87
S. 965 lbs.
1. $9,500
2. 81 hours
3. 11,088 bushels
4. 254 students
5. 7 buses
6. 24 classes
7. 1,950
8. 81,000 gallons
9. 72 seats
10. 2,020

Page 121

1. 37
2. 1 3/4
3. 4 1/12
4. 6

S. 4 yards
S. 28 dollars
1. 1 1/4 gallons
2. 18 girls
3. $98
4. 39 bracelets
5. 2,000 acres
6. 300 miles
7. $48
8. 200 sq. ft.
9. 17 bushels
10. 1 7/20 pounds

Page 122

1. 19.87
2. 2.205
3. .353
4. $2.10

S. $16.90
S. $8.06
1. $10,500
2. $7.14
3. $107.95
4. $7.14
5. $6.90
6. $4.05
7. $12.84
8. $17.11
9. $840
10. $2.62

Page 123

1. 14.4
2. 25%
3. 37.68 ft.
4. 80 ft.

S. 64 seats
S. 12.1
1. $8.95
2. 23 miles
3. 64 boys
4. $18
5. $459
6. 8 pieces; $24
7. $50
8. 154 3/4 pounds
9. $6.72
10. $10.32

Page 124

1. scalene
2. right
3. 117 ft.
4. 256 sq. ft.

S. $32.19
S. 80 ft./$560
1. 480 miles
2. 39 hours 20 minutes
3. 14 students
4. $240
5. 510 students
6. $150
7. $830
8. $1,000,000, 500 lots
9. 200 acres
10. $74.82

Page 125

1. 10,647
2. 60 mph
3. 14 boys
4. 5 pieces
5. 2 7/12 hours
6. 144 miles
7. $.69
8. $3.46
9. 79
10. 2,180 miles
11. 4 1/3 yards
12. 18 girls
13. $4.36
14. $2.75
15. $336
16. $2.76
17. 10 payments
18. 1,200 acres
19. $485.10
20. 272 ft.; $850

Page 126

Final Review

1. 814
2. 2,178
3. 23,965
4. 9,314
5. 7,693
6. 279
7. 3,628
8. 1,414
9. 1,714
10. 5,284
11. 292
12. 28,544
13. 1,665
14. 15,732
15. 158,178
16. 131 r2
17. 344
18. 14 r8
19. 253 r24
20. 69 r1

Page 127

Final Review

1. 5/7
2. 1 1/4
3. 14/15
4. 5 7/8
5. 14 3/10
6. 3/4
7. 3 1/2
8. 2 6/7
9. 5 1/10
10. 4 11/12
11. 6/35
12. 3/44
13. 20
14. 2
15. 8 3/4
16. 2 1/2
17. 4 2/3
18. 1 1/3
19. 3 2/3
20. 4 1/2

Page 128

Final Review

1. 11.913
2. 14.52
3. 30.7
4. 18.6
5. 2.61
6. 70.32
7. 7.92
8. 189.8
9. 1.512
10. .53298
11. 36.5
12. 3,600
13. 2.32
14. 1.34
15. .31
16. 300
17. .03
18. 13
19. .6
20. .35

Page 129

Final Review

1. 13%
2. 3%
3. 70%
4. 19%
5. 60%
6. .08; 2/25
7. .18; 9/50
8. .8; 4/5
9. 2.22
10. 128
11. 64
12. 80%
13. 75%
14. 80%
15. 80%
16. 60%
17. 256 girls
18. 26 games
19. 75% strikes
20. 10% incorrect

Page 130

Final Review

1. answers vary
2. answers vary
3. answers vary
4. answers vary
5. answers vary
6. answers vary
7. answers vary
8. answers vary
9. sides, isosceles, angles, acute
10. sides, scalene angles, obtuse
11. 51 feet
12. 18.84 feet
13. 28.26 sq. ft.
14. 54 sq. ft.
15. 98 sq.ft.
16. 38 1/2 sq. ft.
17. 196 sq. ft.
18. rectangular prism, 12 edges, 6 faces, 8 vertices
19. 96 sq. ft.
20. 148 ft.

Page 131

Final Review

1. -3
2. 3
3. -13
4. -5
5. -24
6. -3
7. 11
8. -12
9. -27
10. -4
11. -32
12. 57
13. 24
14. 36
15. -4
16. 52
17. 17
18. 4
19. -8
20. -8

Page 132

Final Review

1. grade 4
2. 1,500 cans
3. 9,500 cans
4. 21,500 cans
5. grade 2
6. 20 students
7. 11 more C's
8. 100 students
9. 13
10. B's
11. 25 inches
12. 9 inches
13. January
14. 60 inches
15. December
16. 500 boxes
17. Troop 15
18. 200 boxes
19. 900 boxes
20. 750 boxes

Page 133

Final Review

1. 872 miles
2. 435 mph
3. 20 correct
4. 6 bags
5. $2 9/10 = $2.90
6. 150.5 miles
7. $1.56
8. $6.75
9. 98 pounds
10. $290.50
11. 4 1/6 miles
12. 20 correct
13. $6.90
14. $11.19
15. 232 sq. ft.
16. $3.04
17. $600
18. 1/6 of pie
19. $81.15
20. 112 ft.; $1,540

Solutions

Page 134

Review Exercises
1. 1,159
2. 436
3. 2,282

S1. No, members are countable
S2. No, members are uncountable
1. No, 3 is a member of each
2. Yes, can be paired 1-1
3. Answers vary
4. Answers vary
5. {3,5,7,9,11}
6. {0,2,4,6,8,10,12}
7. {3,4,5,6,7,8,9}
8. {10,15,20,25,30}
9. {1,3,5}
10. {8,9,10,11,12}

Problem Solving: 19 girls

Page 135

Review Exercises
1. Answers vary
2. Answers vary
3. Answers vary
4. Answers vary
5. No, 10 is a member of both sets
6. No, cannot be paired in a 1-1 correspondence

S1. Yes, all members of A are members of B
S2. {5,6,7}
1. {1,2,3,4,5,6,7,8}
2. No, not all members of A are members of B
3. {5}, {6}, {7}, {5,6}, {5,7}, {6,7}, {5,6,7}, Ø
4. {1,2,4,5,7}
5. {1,2,4,8}
6. {1,2,3,4,5,6,7}
7. {1,2,3,4,5,6,7,8,10}
8. {1,2,4,6}
9. Yes, can be paired in a 1-1 correspondence
10. No, 6 is common to both sets

Problem Solving: 80 cards

Page 136

Review Exercises
1. {2}
2. {1,2,3,4,6,8}
3. {1,2,3,6}
4. {2}
5. No, cannot be paired in a 1-1 correspondence
6. No, the members are countable

S1. 3
S2. -21
1. 14
2. -18
3. -14
4. -31
5. 40
6. 37
7. -168
8. -13
9. -34
10. -10

Problem Solving: -20°

Page 137

Review Exercises
1. -7
2. 13
3. -39
4. 0
5. A set is a well defined collection of objects
6. A set whose number of members is countable.

S1. -4
S2. -7
1. -2
2. -7
3. -8
4. 14
5. -2
6. 10
7. -22
8. -13
9. -25
10. -84

Problem Solving: $55

Page 138

Review Exercises
1. Ø
2. {0,1,4,5,8,9,10,12,15}
3. {10,15}
4. {1,4,8,9,10,11,12,15}
5. {1,4,8,9,12}
6. Ø

S1. -14
S2. -3
1. 12
2. -3
3. 9
4. -28
5. 46
6. -49
7. -10
8. -11
9. -103
10. 15

Problem Solving: 27 ft.

Page 139

Review Exercises
1. -56
2. 22
3. -35
4. -9
5. -11
6. 4

S1. -48
S2. -126
1. 68
2. -64
3. 288
4. -368
5. -736
6. -24
7. -32
8. 72
9. 72
10. 330

Problem Solving: Floor 31

Page 140

Review Exercises
1. -11
2. -56
3. -2
4. 2
5. -35
6. 36

S1. -4
S2. 6
1. -16
2. 48
3. 15
4. -26
5. -3
6. -2
7. 1
8. -1
9. 3
10. 2

Problem Solving: -11°

Page 141

Review Exercises
1. -2
2. 2
3. -16
4. -1
5. -19
6. -2
7. 13
8. -12
9. -27
10. -1
11. -48
12. 76
13. 56
14. 48
15. -9
16. 42
17. 16
18. 3
19. -2
20. -20

Solutions

Page 142

Review Exercises
1. {1,4}
2. {1,2,3,4,6,7}
3. {1,6}
4. {1,2,3,4,5,6,9}
5. {1,3,4,5,6,7,9}
6. {1,3,4,5,9}

S1. 3/10
S2. -1/10
1. -1/4
2. -1 1/6
3. -2
4. 3/8
5. -7/12
6. 2 1/2
7. 1 1/12
8. 1/8
9. -10
10. 7/15

Problem Solving: 50 sixth graders

Page 143

Review Exercises
1. -59
2. -36
3. -64
4. -28
5. 4
6. 28

S1. -.91
S2. -10.3
1. -12.81
2. 3.17
3. -4.284
4. 12.13
5. 2.37
6. .18
7. .426
8. -6.93
9. 2.01
10. -17.04

Problem Solving: 118 pounds

Page 144

Review Exercises
1. .4
2. -5.9
3. -7.8
4. -5/6
5. -1/10
6. 3/5

S1. 16
S2. -27
1. 216
2. 1
3. 16
4. 32
5. 7
6. 512
7. -1
8. 3,125
9. -125
10. 81

Problem Solving: 2

Page 145

Review Exercises
1. 49
2. 729
3. 36
4. 4
5. 1
6. 9

S1. 12^3
S2. 3^3
1. 2^6
2. $(-9)^3$
3. 16^4
4. 7^2 or $(-7)^2$
5. 10^2 or $(-10)^2$
6. 11^2 or $(-11)^2$
7. $(-1)^4$
8. 2^5
9. 2^4 or 4^2 or $(-2)^4$ or $(-4)^2$
10. 9^6

Problem Solving: (-3)

Page 146

Review Exercises
1. -9
2. -3
3. -7/12
4. 3.6
5. -1.04
6. -3/8

S1. 5
S2. 12
1. 4
2. 11
3. 1
4. 30
5. 10
6. 20
7. 13
8. 3
9. 16
10. 40

Problem Solving: -41

Page 147

Review Exercises
1. 36
2. -8
3. 6^4
4. 8
5. 13
6. 11

S1. 36
S2. 8
1. 2
2. 1
3. 25
4. 88
5. 144
6. 64
7. 54
8. 4
9. 144
10. 9

Problem Solving: 81

Page 148

Review Exercises
1. 13^4
2. 2^7
3. 2^6 or 8^2 or $(-2)^6$ or $(-8)^2$
4. $(-2)^4$
5. 2^3
6. 10^2 or $(-10)^2$
7. 4
8. 8
9. 5
10. 20
11. 3
12. 18
13. 22
14. 4
15. 85
16. 400
17. 63
18. 675
19. 25
20. 10

Page 149

Review Exercises
1. 49
2. 6
3. -2
4. -56
5. -1
6. 24

S1. 28
S2. 38
1. 7
2. 15
3. 34
4. 25
5. 12
6. 27
7. 14
8. 4
9. 28
10. 36

Problem Solving: -1 yard

Solutions

Page 150

Review Exercises
1. {2,4}
2. {8}
3. Ø
4. {1,2,4,5,7,8,9}
5. {1,2,4,5,6,8,10}
6. {1,2,4,5,8}

S1. 4
S2. 9
1. 29
2. 6
3. 2
4. 32
5. 1
6. 7
7. 6
8. 12
9. 7
10. 24

Problem Solving: $29

Page 151

Review Exercises
1. 27
2. 28
3. -13
4. A well-defined collection of objects
5. -1 1/4
6. -.41

S1. commutative (addition)
S2. distributive
1. inverse property (addition)
2. associative (multiplication)
3. identity (addition)
4. inverse (multiplication)
5. commutative (addition)
6. commutative (multiplication)
7. associative (addition)
8. identity (multiplication)
9. distributive
10. inverse (addition)

Problem Solving: 19

Page 152

Review Exercises
1. 11
2. 17
3. 10
4. 23
5. 60
6. 42

S1. $3 + (7 + 9)$
S2. 15×7
1. 1/9
2. $3 \times 6 + 3 \times 2$
3. $12 + 9$
4. $(3 \times 9) \times 5$
5. $3(5 + 7)$
6. -9
7. 1
8. 5
9. $(3 + 5) + 6$
10. $6 \times 4 + 6 \times (-2)$

Problem Solving: $42

Page 153

Review Exercises
1. 10
2. 125
3. 21
4. -90
5. 5
6. -5/6

S1. 2.36×10^9
S2. 1.49×10^{-7}
1. 6.53×10^{11}
2. 1.597×10^5
3. 1.06×10^8
4. 7.216×10^{-6}
5. 1.096×10^9
6. 1.963×10^{-3}
7. 1.6×10^{-10}
8. 8×10^{-10}
9. 7×10^{12}
10. 1.287×10^{-7}

Problem Solving:
1.86×10^5 miles per second

Page 154

Review Exercises
1. 1.23×10^5
2. 3.21×10^{-4}
3. distributive
4. -17
5. -3
6. 35

S1. 7,032,000
S2. .000056
1. 230,000
2. .0000000913
3. .000012362
4. 517,000,000,000
5. 1,127
6. .003012
7. 6,670,000
8. 21,000
9. .00000007
10. 8,000,000

Problem Solving:
93,000,000 miles

Page 155

Review Exercises
1. 1.23×10^5
2. 5.6×10^{-6}
3. 2,760,000
4. .0000375
5. answers vary
6. answers vary

S1. 5/3
S2. 9/2
1. 7/2
2. 6/5
3. 6/5
4. 5/1
5. 6/5
6. 4/3
7. 7/3
8. 3/2
9. 1/2
10. 3/1

Problem Solving: 12/5

Page 156

Review Exercises
1. 2.7×10^{-4}
2. 2.916×10^6
3. 721,000
4. .0000623
5. 30
6. -1.32

S1. yes
S2. no
1. yes
2. no
3. yes
4. yes
5. yes
6. no
7. yes
8. no
9. yes
10. no

Problem Solving: 2

Page 157

Review Exercises
1. 5/3
2. yes ($4 \times 10 = 8 \times 5$)
3. no ($5 \times 5 \neq 2 \times 7$)
4. 16
5. 48
6. 275

S1. 1
S2. 16
1. 3
2. 2
3. 42
4. 6 2/5
5. 6
6. 21
7. 18
8. 2 4/5
9. 1.2
10. 2 1/3

Problem Solving: -32°

Solutions

Page 158

Review Exercises
1. yes (4 × 9 = 3 × 12)
2. n = 30
3. n = 12
4. 2.34×10^8
5. $2.35 \times 10{-}3$
6. 720,000

S1. 204 miles
S2. $12
1. 2 gallons
2. $17.50
3. 15 girls
4. 40 miles
5. 2 4/5 pounds

Problem Solving: 12°

Page 159

Review Exercises
1. 3/1
2. 12/5
3. 8/3
4. yes, 15 × 24 = 12 × 30
5. no, 7 × 9 ≠ 8 × 8
6. yes, 5 × 9 = 3 × 15
7. 4
8. 33
9. 5
10. 35
11. 4 1/2
12. 9
13. 20
14. 3
15. 9
16. $8.40
17. 30 boys
18. 70 miles
19. 2 gallons
20. 4 pounds

Page 160

Review Exercises
1. 21
2. 3 3/5
3. 9
4. 18
5. 26
6. -33

S1. .2, 1/5
S2. .09, 9/100
1. .16, 4/25
2. .06, 3/50
3. .75, 3/4
4. .4, 2/5
5. .01, 1/100
6. .45, 9/20
7. .12, 3/25
8. .05, 1/20
9. .5, 1/2
10. .13, 13/100

Problem Solving: 19/20

Page 161

Review Exercises
1. .8
2. .07
3. 1/4
4. 109.2
5. 128
6. 18

S1. 17.5
S2. 150
1. 4.32
2. 51
3. 15
4. 112.5
5. 32
6. 80
7. 10
8. 216
9. 112.5
10. 13.2

Problem Solving: 34 correct

Page 162

Review Exercises
1. 46.5
2. 24
3. .75
4. 70%
5. 135
6. 600

S1. 25%
S2. 75%
1. 25%
2. 80%
3. 50%
4. 90%
5. 60%
6. 75%
7. 75%
8. 75%
9. 80%
10. 95%

Problem Solving: 75%

Page 163

Review Exercises
1. 3.2
2. 32
3. 75%
4. 90%
5. 50.04
6. 200

S1. 20
S2. 30
1. 48
2. 80
3. 25
4. 4
5. 15
6. 20
7. 60
8. 75
9. 45
10. 125

Problem Solving: 25 students

Page 164

Review Exercises
1. .00072
2. 2 19/10,000
3. 60%
4. 15
5. .021
6. 80

S1. 20 questions
S2. 75%
1. $25
2. $1,600
3. 30
4. 90%
5. 250 cows
6. 25
7. $240
8. 180 boys
9. 20%
10. $210

Problem Solving: $57,600

Page 165

Review Exercises
1. 32.744
2. 2.358
3. 21.98
4. .01248
5. .48
6. 8.1

S1. 30
S2. 30
1. 20%
2. 90%
3. 30 students
4. 18 passes
5. 20% are red
6. 75%
7. 250 students
8. $25
9. 60%
10. $691.20

Problem Solving: 97

Solutions

Page 166

Review Exercises
1. 13%
2. 3%
3. 70%
4. 19%
5. 60%
6. .08, 2/25
7. .18, 9/50
8. .8, 4/5
9. 2.22
10. 128
11. 80%
12. 75%
13. 12
14. 75
15. 80%
16. 60%
17. 256 girls
18. 26 games
19. 75%
20. 150 students

Page 167

Review Exercises
1. 70%
2. 80%
3. 3/25
4. 12
5. 25%
6. 25

S1. 1, 30, 2, 15, 3, 10, 5, 6
S2. 1, 36, 2, 18, 3, 12, 4, 9, 6
1. 1, 100, 2, 50, 4, 25, 5, 20, 10
2. 1, 42, 2, 21, 3, 14, 6,7
3. 1, 70, 2, 35, 5, 14, 7, 10
4. 1, 81, 3, 27, 9
5. 1, 50, 2, 25, 5, 10
6. 1, 40, 2, 20, 4, 10, 5, 8
7. 1, 75, 3, 25, 5, 15
8. 1, 90, 2, 45, 3, 30, 5, 18, 6, 15, 9, 10
9. 1, 20, 2, 10, 4, 5
10. 1, 28, 2, 14, 4, 7

Problem Solving: 54 correct

Page 168

Review Exercises
1. -18
2. 24
3. 20
4. 7 1/2
5. 15
6. 25%

S1. 2
S2. 4
1. 2
2. 3
3. 14
4. 16
5. 20
6. 10
7. 5
8. 12
9. 12
10. 20

Problem Solving:
5,879,000,000,000 miles

Page 169

Review Exercises
1. 1.2×10^{-6}
2. 4.96×10^{8}
3. 13,200,000
4. .00000464
5. 1,60,2,30,3,20,4,15,5,12,6,10
6. 4

S1. 4, 6, 8, 10
S2. 0, 12, 18, 30
1. 10, 15, 20, 25
2. 0, 6, 12, 15
3. 0, 30, 40, 50
4. 0, 4, 8
5. 22, 44
6. 24, 32, 40
7. 60, 80, 100
8. 14, 28, 35
9. 90, 120, 150
10. 27, 45

Problem Solving: 120 pitches

Page 170

Review Exercises
1. 1, 30, 2, 15, 3, 10, 5, 6
2. 4
3. 0, 8, 16, 24, 32, 40
4. 3
5. 25%
6. 35

S1. 12
S2. 24
1. 15
2. 30
3. 60
4. 30
5. 36
6. 60
7. 48
8. 40
9. 36
10. 60

Problem Solving: $12.96

Page 171

1. 1, 24, 2, 12, 3, 8, 4, 6
2. 1, 16, 2, 8, 4
3. 1, 32, 2, 16, 4, 8
4. 1, 28, 2, 14, 4, 7
5. 1, 70, 2, 35, 5, 14, 7, 10
6. 1, 25, 5
7. 4
8. 12
9. 20
10. 5
11. 7
12. 18
13. 9, 12, 15
14. 9, 27, 36, 45
15. 15, 30, 45, 60
16. 12
17. 60
18. 60
19. 12
20. 24

Page 172

Review Exercises
1. 8
2. 12
3. 1, 28, 2, 14, 4, 7
4. 0, 12, 24, 36, 48, 60
5. yes, $3 \times 8 = 4 \times 6$
6. 9

S1. 0
S2. -7, -2, 10
1. 5, 9
2. 3, 4
3. 2, 4, 7
4. -5, -8
5. 10, 9, -6, 6
6. 9, -7, 1
7. -8, 8, 0, -3
8. 0, 7, 8
9. -6, 4, -3
10. 5, -3, 9, -8

Problem Solving: 8°

Page 173

Review Exercises
1. {4,6,8,10}
2. {1,3,4,5,6,8,9,10}
3. {2,4,5,6,8,9,10}
4. {4,5,6,8,10}
5. no, cannot be paired in 1-1 correspondence
6. no, they have members in common

S1. 1
S2. 7
1. 3
2. 5
3. 5
4. 18
5. 4
6. 11
7. 8
8. 7
9. 4
10. 12

Problem Solving: 512 miles

Page 174

Review Exercises
1. 7
2. -22
3. 8
4. -42
5. 5
6. -2

S1. (2, 1)
S2. (-4, -2)
1. (6, 3)
2. (2, -5)
3. (-7, -3)
4. (-5, 1)
5. (4, 6)
6. (4, -3)
7. (-3, -5)
8. (-2, 2)
9. (2, 1)
10. (-6, 7)

Problem Solving: $24

Page 175

Review Exercises
1. -7/15
2. -.68
3. 3/8
4. 1
5. -5
6. 2

S1. B
S2. A
1. C
2. D
3. F
4. M
5. E
6. J
7. H
8. G
9. K
10. I

Problem Solving: 95%

Page 176

Review Exercises
1. 32
2. 22
3. 5
4. 72
5. 1.7×10^{-4}
6. 2.13×10^{5}

S1. 1/3
S2. 3/2
1. -3/2
2. 1/3
3. -2
4. 5/8
5. 1/2
6. 7/2
7. 6/5
8. 4/5
9. 3/2
10. -3/2

Problem Solving: 320 girls

Page 177

1. -4
2. 9, 0, 10
3. 3, 2, -7
4. 8, 1, -4, -8
5. 3
6. 6
7. 5
8. 8
9. (5, 3)
10. (-6, 1)
11. (4, -4)
12. (-6, -5)
13. (2, 6)
14. 5/3
15. B
16. I
17. H
18. D
19. A
20. 1/8

Page 178

Review Exercises
1. 6
2. -18
3. -24
4. -16
5. -7
6. -8

Problem Solving: $6.00

S1.
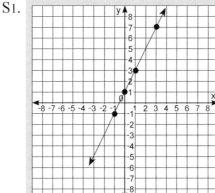

S2.
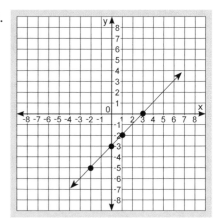

1.
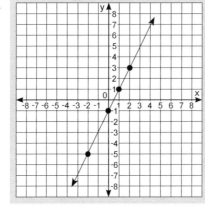

2.
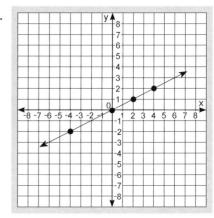

Page 179

Review Exercises
1. 8 2/5
2. 3
3. 75
4. -17/24
5. -3
6. -2.1

Problem Solving: 80%

S1.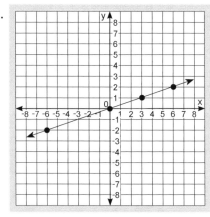

S2.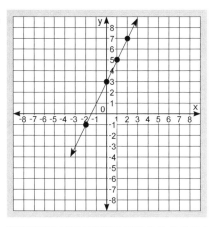

1.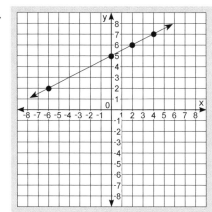

2.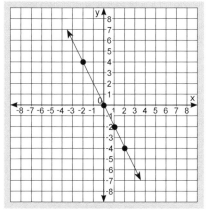

Page 180	**Page 181**	**Page 182**	**Page 183**
Review Exercises	Review Exercises	Review Exercises	Review Exercises
1. 1.72×10^6	1. 10	1. 1	1. 1/10
2. 3.8×10^{-7}	2. 23	2. 72	2. -4/5
3. 196,300,000	3. 102	3. 2	3. -1 1/3
4. .00034	4. -8	4. 2.1×10^5	4. -3
5. -10	5. 6	5. 3.16×10^{-3}	5. -.64
6. .19	6. 21	6. 16	6. -1.1
S1. 5	S1. 12	S1. 7	S1. 3
S2. 32	S2. 3	S2. 4	S2. 45
1. -3	1. -14	1. -1	1. 8
2. 40	2. -12	2. -4	2. -3
3. -13	3. -9	3. -15	3. 10
4. -5	4. -5	4. 75	4. -3
5. 13	5. 18	5. 12	5. 5
6. -12	6. -6	6. 1	6. -4
7. 34	7. 2	7. 2	7. -10
8. 4	8. 20	8. 50	8. -7
9. -11	9. -15	9. 10	9. -3
10. -27	10. 14	10. -3	10. 10
Problem Solving: 7/3	Problem Solving: 9/4	Problem Solving: 200 students	Problem Solving: 14

Solutions

Page 184

Review Exercises
1. 216
2. 1
3. 9
4. 13
5. 12
6. 158

S1. -1
S2. 3
1. 7
2. 2
3. 4
4. 4
5. 3
6. 1
7. -1
8. 4
9. 2
10. 6

Problem Solving: 20%

Page 185

Review Exercises
1. 20
2. no, $4 \times 5 \neq 3 \times 7$
3. 8/3
4. 60
5. 25%
6. 35

S1. 4
S2. 15
1. -2
2. -3
3. 10
4. -5
5. 12
6. 1
7. 7
8. 3
9. -25
10. -10

Problem Solving:
8 blue marbles

Page 186

Review Exercises
1. 1, 48, 2, 24, 3, 16, 4, 12, 6, 8
2. 8
3. 30
4. 5
5. 20
6. -5

S1. $2x - 7 = 12$
S2. $3x + 2 = 30$
1. $2x + 5 = 14$
2. $4x - 6 = 10$
3. $4x - 5 = 12$
4. $x/3 - 4 = 2x + 8$
5. $2(x + 2) = 10$
6. $5x - 3 = 17$
7. $2x - 6 = 15$
8. $3x - 2 = 2x + 7$
9. $x + 4 = 7 + -12$
10. $n/5 = 25$

Problem Solving: 227.5 miles

Page 187

Review Exercises
1. 7
2. -4
3. 12
4. 5
5. 15
6. 3

S1. $2x - 6 = 16, x = 11$
S2. $3x - 8 = 28, x = 12$
1. $2x - 5 = 67, x = 36$
2. $4x - 5 = -17, x = -3$
3. $4x - 6 = 2x + 8, x = 7$
4. $x/2 + 8 = 10, x = 4$
5. $4x - 2 = 10, x = 3$

Problem Solving: $12,500

Page 188

Review Exercises
1. .000000361
2. 1.27×10^{-6}
3. 7.29×10^{8}

S1. Kevin is 12
    Amir is 18
S2. Short piece = 11 inches
    Long piece = 33 inches
1. Bob earned $20
    Bill earned $46
2. Monday, $91
    Tuesday, $121
3. 12 years old
4. Weekly salary is $140
5. John is 30

Problem Solving: 89

Page 189

Review Exercises
1. -11
2. 20
3. 44
4. -9
5. 3
6. 11

S1. 6
S2. Bart is 110 pounds
    Steve is 160 pounds
1. -9
2. 6
3. -24
4. Ellen is 11
    Roy is 33
5. 3

Problem Solving:
600 miles per hour

Page 190

1. -7
2. 13
3. 56
4. 3
5. -5
6. -18
7. -7
8. 15
9. 8
10. 8
11. -11
12. 21
13. 12
14. -5
15. 6
16. 6
17. Sue, $22
    Ann, $44
18. Ron is 28
    Bill is 36
19. -2
20. 18

Page 191

Review Exercises
1. 0,8,16,24,32,40,48
2. 1, 60, 2, 30, 3, 20, 4, 15, 5, 12, 6, 10
3. 20
4. 6×10^{-6}
5. 2.1×10^{6}
6. .0021

S1. 3/12 = 1/4
S2. 7/12
1. 2/12 = 1/6
2. 1/12
3. 9/12 = 3/4
4. 10/12 = 5/6
5. 9/12 = 3/4
6. 5/12
7. 10/12 = 5/6
8. 6/12 = 1/2
9. 6/12 = 1/2
10. 4/12 = 1/3

Problem Solving: -3

Solutions

Page 192

Review Exercises
1. -10
2. -25
3. -10
4. 4
5. 15
6. 9

S1. 1/8
S2. 4/8 = 1/2
1. 1/8
2. 7/8
3. 4/8 = 1/2
4. 4/8 = 1/2
5. 2/8 = 1/4
6. 0/8
7. 2/8 = 1/4
8. 5/8
9. 5/8
10. 4/8 = 1/2

Problem Solving: 20 pounds

Page 193

Review Exercises
1. 30%
2. 3%
3. 60%
4. 2
5. 25%
6. 20

S1. range 7, mode 4
S2. range 6, mode 6
1. range 6, mode 7
2. range 23, mode 30
3. range 10, mode 3, 6
4. range 19, mode 9
5. range 7, mode 3
6. range 11, mode 8
7. range 9, mode 2
8. range 6, mode 91
9. range 9, mode 2
10. range 19, mode 2

Problem Solving: $.38

Page 194

Review Exercises
1. 8/5
2. no, $7 \times 11 \neq 8 \times 9$
3. 20
4. 1.28×10^6
5. 9.62×10^{-5}
6. .000062

S1. mean 3, median 3
S2. mean 4, median 4
1. mean 3, median 2
2. mean 3, median 2
3. mean 15, median 15
4. mean 2, median 1
5. mean 6, median 6
6. mean 124, median 126
7. mean 4, median 4
8. mean 4, median 4
9. mean 5, median 4
10. mean 50, median 50

Problem Solving: 32 students

Page 195

Review Exercises
1. 21
2. 14
3. 15
4. 15
5. 84
6. 11

S1. 6
S2. 2
1. 5
2. 6
3. 3
4. 2
5. 8
6. 4
7. 10
8. 2
9. 5
10. 5

Problem Solving:
186,000 miles per second

Page 196

1. 3/10
2. 4/10 = 2/5
3. 5/10 = 1/2
4. 7/10
5. 8/10 = 4/5
6. 6/10 = 3/5
7. 1/8
8. 4/8 = 1/2
9. 5/8
10. 2/8 = 1/4
11. 4/8 = 1/2
12. 2/8 = 1/4
13. 8
14. 4
15. 5
16. 4
17. 2
18. 4
19. 7
20. 3

Page 197

1) -7
2) 7
3) -11
4) -101
5) 7
6) -284
7) -160
8) -981
9) -1,540
10) 454
11) -95
12) 15
13) 11
14) -458
15) 500

Page 198

Exercises

1) -16
2) -5
3) -3
4) 11
5) 9
6) -35
7) 49
8) -168
9) 27
10) 6
11) 2
12) -24
13) -202
14) -172
15) 537
16) 106
17) 1
18) -945
19) -100
20) 36

Page 199

Exercises

1) -231
2) -276
3) 152
4) -368
5) 384
6) 441
7) -288
8) -600
9) -312
10) 612
11) -36
12) 72
13) 360
14) -180
15) 72
16) 216
17) 24
18) 168
19) -24
20) -504

Page 200

Exercises

1) -8
2) 16
3) -18
4) -64
5) 170
6) -14
7) -9
8) -23
9) 2
10) 24
11) 24
12) 48
13) 3
14) -30
15) -1
16) 3
17) -1
18) 15
19) 7
20) -2

Page 201

Exercises

1) $\frac{1}{6}$
2) $\frac{7}{20}$
3) $-1\frac{1}{12}$
4) $2\frac{1}{4}$
5) $-\frac{1}{3}$
6) $-\frac{7}{20}$
7) $-\frac{9}{10}$
8) $\frac{1}{6}$
9) $-\frac{3}{20}$
10) $-1\frac{2}{15}$
11) $1\frac{1}{8}$
12) $-1\frac{1}{4}$
13) $-\frac{1}{4}$
14) $2\frac{1}{24}$
15) $-4\frac{1}{2}$
16) 2
17) -7
18) $2\frac{1}{4}$
19) $\frac{1}{10}$
20) $-\frac{11}{12}$

Page 202

Exercises

1) - 2.36
2) -12.69
3) 1.28
4) 5.3562
5) -8.27
6) -1.53
7) -29.28
8) -2.67
9) -1.6
10) -4.338
11) -13
12) 1.344
13) 10.58
14) -3.195
15) -6.384
16) -3.11
17) -27.04
18) 1.05
19) -7.87
20) 12.5

Page 203

Exercises

1) 81	17) 10^2
2) 9	18) 50^2
3) -216	19) $(-10)^4$
4) 125	20) 7^2
5) 2,401	
6) 81	
7) -512	
8) 6,561	
9) 256	
10) -343	
11) 5^5	
12) 6^2	
13) $(-2)^4$	
14) 12^2	
15) 15^2	
16) $(-9)^3$	

Page 204

Exercises

1) 6^8	17) 3^6
2) $\dfrac{8^3}{5^3}$	18) $\dfrac{1}{5^3}$
3) $4^3 \times 6^3$	19) 5^9
4) 5^{12}	20) 6^3
5) 3^5	
6) $\dfrac{1}{6^4}$	
7) 7^8	
8) $3^3 \cdot 9^3$	
9) 4^{10}	
10) $\dfrac{1}{3^3}$	
11) 6	
12) 1	
13) $\dfrac{1}{4^2}$	
14) 5^9	
15) $\dfrac{1}{5^3}$	
16) $\dfrac{7^4}{8^4}$	

Page 205

Exercises

1) 6	17) $5\sqrt{5}$
2) 11	18) 70
3) 50	19) 3
4) 20	20) $\dfrac{5}{6}$
5) $5\sqrt{3}$	
6) $2\sqrt{2}$	
7) 4	
8) 60	
9) $\dfrac{\sqrt{5}}{3}$	
10) $\dfrac{2\sqrt{3}}{5}$	
11) $5\sqrt{2}$	
12) $6\sqrt{2}$	
13) $2\sqrt{3}$	
14) $10\sqrt{2}$	
15) 2	
16) $3\sqrt{3}$	

Page 206

Exercises

1) 31	17) 50
2) 38	18) -3
3) 72	19) 29
4) 48	20) 45
5) 45	
6) -33	
7) 83	
8) 17	
9) 3	
10) 2	
11) 66	
12) 317	
13) 6	
14) 45	
15) 50	
16) 2	

Page 207

Exercises

1) distributive
2) identity of +
3) commutative of +
4) commutative of x
5) inverse of +
6) inverse of x
7) associative of +
8) identity of x
9) inverse of +
10) identity of +
11) commutative of x
12) identity of x
13) distributive
14) commutative of +
15) associative of x
16) inverse of +
17) distributive
18) commutative of x
19) inverse of +
20) commutative of +

Page 208

Exercises

1) -2	17) -103, 100, 180
2) 10, 1, -3	18) $\dfrac{1}{4}, \dfrac{1}{2}, \dfrac{3}{4}$
3) -6, 3, 5	19) $1\dfrac{7}{8}, 2, 3\dfrac{1}{2}$
4) 7, 5, 8, -1	20) -15, -7, 9
5) 9, 7	
6) -2, -7, -6, -8	
7) 7, 6, 10, -2	
8) -6, 5, 4, 7	
9) -4, 5, -2, -5	
10) -15, 5, 20	
11) -5, $\dfrac{1}{5}$, 15	
12) $\dfrac{3}{5}, 1, \dfrac{7}{5}$	
13) $\dfrac{4}{5}, \dfrac{5}{4}, \dfrac{5}{1}$	
14) $2^3, 5^2, 3^3$	
15) $2\dfrac{1}{8}, 3\dfrac{1}{3}, 4\dfrac{1}{2}$	
16) $\dfrac{3}{4}, 1, \dfrac{8}{5}$	

Solutions

Page 209

Exercises

1) (-5, 4)
2) (5, 3)
3) (-3, -5)
4) (1, -2)
5) (0, 4)
6) (-4, 2)
7) (2, -4)
8) (-3, -3)
9) (-2, 3)
10) (5, -4)
11) P
12) M
13) O
14) R
15) J
16) N
17) F
18) Q
19) B
20) L

Page 210

Exercises

1) no - 7 is paired with two y-values
2) yes - each x-value is paired with exactly one y-value
3) 2, 4, 7
4) 3, 5, 8, 9
5) 1, 3, 4
6) 3, 5, 7
7) A vertical line can only intersect a graph at one point. The graph is a function.
8) False
9) True
10) 1, 3, 4, 6
11) 2, 5, 9
12) yes - each x-value is paired with exactly one y-value
13) 3, 5, 7
14) 4, 6, 8, 9
15) no - 5 is paired with two y-values

Page 211

Exercises

1) $2 \times 3 \times 7$
2) $2^2 \times 3^2$
3) 2×3^3
4) $2 \times 5 \times 31$
5) 5^3
6) $2 \times 3 \times 107$
7) 3×31
8) 2×7^2
9) $3 \times 5 \times 7$
10) $2^5 \times 3$
11) $2^4 \times 3^2$
12) $2^3 \times 5^2$
13) $3^3 \times 5$
14) $2^3 \times 3^3$
15) $2^2 \times 3 \times 5$
16) $2^4 \times 3 \times 5$
17) Prime
18) $2 \times 3^2 \times 5$
19) $2 \times 3 \times 17$
20) $2^2 \times 3^2 \times 5$

Page 212

Exercises

1) 5
2) 4
3) 7
4) 6
5) 25
6) 6
7) 2
8) 8
9) 12
10) 11
11) 45
12) 14
13) 3
14) 22
15) 6
16) 105
17) 14
18) 6
19) 21
20) 18

Page 213

Exercises

1) 60
2) 60
3) 66
4) 70
5) 144
6) 210
7) 120
8) 225
9) 196
10) 330
11) 36
12) 180
13) 30
14) 24
15) 60
16) 120
17) 600
18) 240
19) 1,400
20) 720

Solutions

Page 214

Exercises
1) 1.449×10^{10}
2) 3.46×10^{-5}
3) 2.5979×10^5
4) 7.21×10^{-6}
5) 1.079×10^9
6) 7.6×10^{-3}
7) 1.9×10^{-7}
8) 2.4×10^5
9) 7.62×10^{-5}
10) 5.4×10^7
11) 609,300
12) 230,000
13) .000006
14) 13,470
15) 321,000
16) .0042
17) 450,000
18) 5,000,000
19) .00372
20) 3,950,000

Page 215

Exercises
1) $3\frac{3}{4}$
2) $1\frac{5}{7}$
3) $3\frac{1}{3}$
4) $\frac{3}{4}$
5) 20
6) 12
7) $13\frac{1}{2}$
8) $3\frac{3}{5}$
9) $6\frac{1}{4}$
10) 32
11) $6\frac{1}{9}$
12) $17\frac{1}{2}$
13) 63
14) $15\frac{3}{4}$
15) 50
16) 2
17) 14
18) 35
19) $2\frac{1}{7}$
20) $6\frac{2}{9}$

Page 216

Exercises
1) $154
2) 5 cans
3) $27^1/_2$ miles
4) 360 miles
5) 40 liters
6) 80 cups of flour
7) $2.00
8) $2^1/_2$ cups
9) $1^3/_4$ inch
10) 35 hours
11) 700 boys
12) $1.65
13) $^4/_{15}$ minutes = 16 seconds
14) $4^1/_4$ inches
15) 25 bags of cement

Page 217

Exercises
1) 42
2) 21
3) 25%
4) 25
5) 20
6) 60
7) 75%
8) 16
9) 25%
10) 20%
11) 5%
12) 25.6
13) 80
14) 24
15) 24
16) 75%
17) 6
18) 320
19) 25%
20) 202.5

Page 218

Exercises
1) 30
2) 30
3) 20%
4) 90%
5) 24 absent
6) 18 caught
7) 20% are red
8) 75% won
9) 250 in the school
10) $25
11) 60%
12) $691.20
13) 75% correct
14) $936
15) 400 enrolled
16) $500
17) 54 correct
18) 75% complete
19) 528 present
20) 50 students

Page 219
Final Review

1. {2,3,4}
2. {0,1,2,3,4,5,6,8,9}
3. {1,2,4,5}
4. 3
5. -23
6. 36
7. 8
8. -.55
9. -9/10
10. 125
11. 7
12. 33
13. 22
14. 68
15. 2
16. 20
17. commutative (addition)
18. distributive
19. 1.28×10^9
20. 6.53×10^{-6}

Page 220
Final Review

21. 60,900,000
22. .00000762
23. 9/5
24. no, $8 \times 3 \neq 4 \times 7$
25. 21
26. 6 blue marbles
27. 4
28. 25%
29. 32
30. 24
31. 40%
32. 30
33. 1, 40, 2, 20, 4, 10, 5, 8
34. 20
35. 0, 6, 12, 18, 24, 30, 36
36. 24
37. -8, -2, 0
38. -5, 9, -1, -7
39. -6, 4, -3
40. 9, 8, 1

Page 221
Final Review

41. (6, -3)
42. (-5, 1)
43. (4, -3)
44. (-4, -2)
45. (4, 6)
46. H
47. G
48. L
49. J
50. K
51. 2/3
52. 3/8

53.

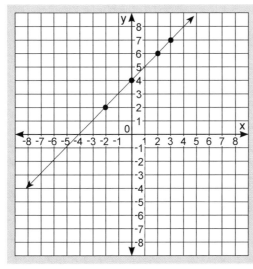

54.

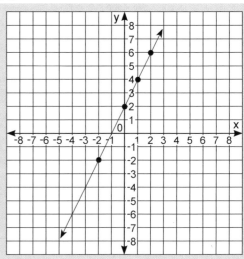

Page 222
Final Review

55. 9
56. -15
57. 18
58. -3
59. 6
60. -3
61. 24
62. 4
63. -6
64. 1
65. 9
66. 12
67. 50
68. Jane, $16
    Sue, $48
69. Maria is 22
    Al is 29
70. 9

Page 223
Final Review

71. 5/16
72. 6/16 = 3/8
73. 7/16
74. 11/16
75. 1/8
76. 4/8 = 1/2
77. 4/8 = 1/2
78. 3/8
79. 8
80. 3
81. 5
82. 3
83. 4
84. 8
85. 14
86. 6

Page 224
Final Review

87. 1 1/4
88. 3 1/2
89. 5 7/10
90. 2
91. 8 3/4
92. 4 2/3
93. 3 2/3
94. 19.42
95. 5.025
96. 70.32
97. 15.33
98. .432
99. 1.34
100. .03

Glossary

absolute value   The distance of a number from 0 on the number line. The absolute value is always positive.

acute angle   An angle with a measure of less than 90 degrees.

adjacent   Next to.

algebraic expression   A mathematical expression that contains at least one variable.

angle   Any two rays that share an endpoint will form an angle.

associative properties   For any a, b, c:
addition: (a + b) + c = a + (b + c)
multiplication: (ab)c = a(bc)

base   The number being multiplied. In an exponential expression such as 4^2, 4 is the base.

coefficient   A number that multiplies the variable. In the term 7x, 7 is the coefficient of x.

commutative properties   For any a, b:
addition: a + b = b + a
multiplication: ab = ba

complementary angles   Two angles that have measures whose sum is 90 degrees.

congruent   Two figures having exactly the same size and shape.

coordinate plane   The plane which contains the x- and y-axes. It is divided into 4 quadrants. Also called coordinate system and coordinate grid.

coordinates   An ordered pair of numbers that identify a point on a coordinate plane.

data   Information that is organized for analysis.

degree   A unit that is used in measuring angles.

denominator   The bottom number of a fraction that tells the number of equal parts into which a whole is divided.

disjoint sets   Sets that have no members in common. {1,2,3} and {4,5,6} are disjoint sets.

Glossary

distributive property   For real numbers a, b, and c: a(b + c) = ab + ac.

E

element of a set   Member of a set.

empty set   The set that has no members. Also called the null set and written Ø or { }.

equation   A mathematical sentence that contains an equal sign (=) and states that one expression is equal to another expression.

equivalent   Having the same value.

exponent   A number that indicates the number of times a given base is used as a factor. In the expression n^2, 2 is the exponent.

expression   Variables, numbers, and symbols that show a mathematical relationship.

extremes of a proportion   In the proportion $\frac{a}{b} = \frac{c}{d}$, a and d are the extremes.

F

factor   An integer that divides evenly into another.

finite   Something that is countable.

formula   A general mathematical statement or rule. Used often in algebra and geometry.

function   A set of ordered pairs that pairs each x-value with one and only one y-value. (0,2), (-1,6), (4,-2), (-3,4) is a function.

G

graph   To show points named by numbers or ordered pairs on a number line or coordinate plane. Also, a drawing to show the relationship between sets of data.

greatest common factor   The largest common factor of two or more numbers. Also written GCF. The greatest common factor of 15 and 25 is 5.

grouping symbols   Symbols that indicate the order in which mathematical operations should take place. Examples include parentheses ( ), brackets [ ], braces { }, and fraction bars — .

H

hypotenuse   The side opposite the right angle in a right triangle.

I

identity properties of addition and multiplication   For any real number a:
addition: a + 0 = 0 + a = a
multiplication: $1 \times a = a \times 1 = a$

inequality   A mathematical sentence that states one expression is greater than or less than another. Inequality symbols are read as follows: < less than
≤ less than or equal to
> greater than
≥ greater than or equal to

infinite   Having no boundaries or limits. Uncountable.

integers   Numbers in a set. {...-3, -2, -1, 0, 1, 2, 3...}

intersection of sets   If A and B are sets, then A intersection B is the set whose members are included in both sets A and B, and is written A ∩ B. If set A = {1,2,3,4} and set B = {1,3,5}, then A ∩ B = {1,3}

inverse properties of addition and multiplication   For any number a:
addition: a + -a = 0
multiplication: $a \times 1/a = 1$ (a ≠ 0)

inverse operations   Operations that "undo" each other. Addition and subtraction are inverse operations, and multiplication and division are inverse operations.

L

least common multiple   The least common multiple of two or more whole numbers is the smallest whole number, other than zero, that they all divide into evenly. Also written as LCM. The least common multiple of 12 and 8 is 24.

linear equation   An equation whose graph is a straight line.

M

mean   In statistics, the sum of a set of numbers divided by the number of elements in the set. Sometimes referred to as average.

means of a proportion   In the proportion $\frac{a}{b} = \frac{c}{d}$, b and c are the means.

median   In statistics, the middle number of a set of numbers when the numbers are arranged in order of least to greatest. If there are two middle numbers, find their mean.

mode   In statistics, the number that appears most frequently. Sometimes there is no mode. There may also be more than one mode.

multiple   The product of a whole number and another whole number.

Glossary

natural numbers   Numbers in the set {1, 2, 3, 4, ...}. Also called counting numbers.

negative numbers   Numbers that are less than zero.

null set   The set that has no members. Also called the empty set and written Ø or { }.

number line   A line that represents numbers as points.

numerator   The top part of a fraction.

obtuse angle   An angle whose measure is greater than 90° and less than 180°.

opposites   Numbers that are the same distance from zero, but are on opposite sides of zero on a number line. 4 and -4 are opposites.

order of operations   The order of steps to be used when simplifying expressions.
1. Evaluate within grouping symbols.
2. Eliminate all exponents.
3. Multiply and divide in order from left to right.
4. Add and subtract in order from left to right.

ordered pair   A pair of numbers (x,y) that represent a point on the coordinate plane. The first number is the x-coordinate and the second number is the y-coordinate.

origin   The point where the x-axis and the y-axis intersect in a coordinate plane. Written as (0,0).

outcome   One of the possible events in a probability situation.

parallel lines   Lines in a plane that do not intersect. They stay the same distance apart.

percent   Hundredths or per hundred. Written %.

perimeter   The distance around a figure.

perpendicular lines   Lines in the same plane that intersect at a right (90°) angle.

pi   The ratio of the circumference of a circle to its diameter. Written π. The approximate value for π is 3.14 as a decimal and $\frac{22}{7}$ as a fraction.

plane   A flat surface that extends infinitely in all directions.

point   An exact position in space. Points also represent numbers on a number line or coordinate plane.

positive number   Any number that is greater than 0.

power   An exponent.

prime number   A whole number greater than 1 whose only factors are 1 and itself.

probability   What chance, or how likely it is for an event to occur. It is the ratio of the ways a certain outcome can occur and the number of possible outcomes.

proportion   An equation that states that two ratios are equal. $\frac{4}{8} = \frac{2}{4}$ is a proportion.

Pythagorean theorem   In a right triangle, if c is the hypotenuse, and a and b are the other two legs, then $a^2 + b^2 = c^2$.

Q

quadrant   One of the four regions into which the x-axis and y-axis divide a coordinate plane.

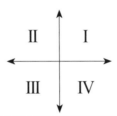

R

range   The difference between the greatest number and the least number in a set of numbers.

ratio   A comparison of two numbers using division. Written a:b, a to b, and a/b.

reciprocals   Two numbers whose product is 1. $\frac{2}{3}$ and $\frac{3}{2}$ are reciprocals because $\frac{2}{3} \times \frac{3}{2} = 1$.

reduce   To express a fraction in its lowest terms.

relation   Any set of ordered pairs.

right angle   An angle that has a measure of 90°.

rise   The change in y going from one point to another on a coordinate plane. The vertical change.

run   The change in x going from one point to another on a coordinate plane. The horizontal change.

S

scientific notation   A number written as the product of a numbers between 1 and 10 and a power of ten. In scientific notation, $7,000 = 7 \times 10^3$.

set   A well-defined collection of objects.

slope   Refers to the slant of a line. It is the ratio of rise to run.

Glossary

solution A number that can be substituted for a variable to make an equation true.

square root Written $\sqrt{}$. The $\sqrt{36} = 6$ because $6 \times 6 = 36$.

statistics Involves data that is gathered about people or things and is used for analysis.

subset If all the members of set A are members of set B, then set A is a subset of set B. Written $A \subset B$. If set A = {1,2,3} and set B = {0,1,2,3,5,8}, set A is a subset of set B because all of the members of a set A are also members of set B.

U

union of sets If A and B are sets, the union of set A and set B is the set whose members are included in set A, or set B, or both set A and set B. A union B is written $A \cup B$. If set = {1,2,3,4} and set B = {1,3,5,7}, then $A \cup B$ = {1,2,3,4,5,7}.

universal set The set which contains all the other sets which are under consideration.

V

variable A letter that represents a number.

Venn diagram A type of diagram that shows how certain sets are related.

vertex The point at which two lines, line segments, or rays meet to form an angle.

W

whole number Any number in the set {0, 1, 2, 3, 4...}

X

x-axis The horizontal axis on a coordinate plane.

x-coordinate The first number in an ordered pair. Also called the abscissa.

Y

y-axis The vertical axis on a coordinate plane.

y-coordinate The second number in an ordered pair. Also called the ordinate.

Important Symbols

<	less than		π	pi
≤	less than or equal to		{ }	set
>	greater than		| |	absolute value
≥	greater than or equal to		$.\overline{n}$	repeating decimal symbol
=	equal to		1/a	the reciprocal of a number
≠	not equal to		%	percent
≅	congruent to		(x,y)	ordered pair
( )	parenthesis		⊥	perpendicular
[ ]	brackets		| |	parallel to
{ }	braces		∠	angle
...	and so on		∈	element of
• or ×	multiply		∉	not an element of
∞	infinity		∩	intersection
a^n	the nth power of a number		∪	union
$\sqrt{}$	square root		⊂	subset of
Ø, { }	the empty set or null set		⊄	not a subset of
∴	therefore		△	triangle
°	degree			

Multiplication Table

x	2	3	4	5	6	7	8	9	10	11	12
2	4	6	8	10	12	14	16	18	20	22	24
3	6	9	12	15	18	21	24	27	30	33	36
4	8	12	16	20	24	28	32	36	40	44	48
5	10	15	20	25	30	35	40	45	50	55	60
6	12	18	24	30	36	42	48	54	60	66	72
7	14	21	28	35	42	49	56	63	70	77	84
8	16	24	32	40	48	56	64	72	80	88	96
9	18	27	36	45	54	63	72	81	90	99	108
10	20	30	40	50	60	70	80	90	100	110	120
11	22	33	44	55	66	77	88	99	110	121	132
12	24	36	48	60	72	84	96	108	120	132	144

Commonly Used Prime Numbers

2	3	5	7	11	13	17	19	23	29
31	37	41	43	47	53	59	61	67	71
73	79	83	89	97	101	103	107	109	113
127	131	137	139	149	151	157	163	167	173
179	181	191	193	197	199	211	223	227	229
233	239	241	251	257	263	269	271	277	281
283	293	307	311	313	317	331	337	347	349
353	359	367	373	379	383	389	397	401	409
419	421	431	433	439	443	449	457	461	463
467	479	487	491	499	503	509	521	523	541
547	557	563	569	571	577	587	593	599	601
607	613	617	619	631	641	643	647	653	659
661	673	677	683	691	701	709	719	727	733
739	743	751	757	761	769	773	787	797	809
811	821	823	827	829	839	853	857	859	863
877	881	883	887	907	911	919	929	937	941
947	953	967	971	977	983	991	997	1009	1013

Squares and Square Roots

No.	Square	Square Root	No.	Square	Square Root	No.	Square	Square Root
1	1	1.000	51	2,601	7.141	101	10201	10.050
2	4	1.414	52	2,704	7.211	102	10,404	10.100
3	9	1.732	53	2,809	7.280	103	10,609	10.149
4	16	2.000	54	2,916	7.348	104	10,816	10.198
5	25	2.236	55	3,025	7.416	105	11,025	10.247
6	36	2.449	56	3,136	7.483	106	11,236	10.296
7	49	2.646	57	3,249	7.550	107	11,449	10.344
8	64	2.828	58	3,364	7.616	108	11,664	10.392
9	81	3.000	59	3,481	7.681	109	11,881	10.440
10	100	3.162	60	3,600	7.746	110	12,100	10.488
11	121	3.317	61	3,721	7.810	111	12,321	10.536
12	144	3.464	62	3,844	7.874	112	12,544	10.583
13	169	3.606	63	3,969	7.937	113	12,769	10.630
14	196	3.742	64	4,096	8.000	114	12,996	10.677
15	225	3.873	65	4,225	8.062	115	13,225	10.724
16	256	4.000	66	4,356	8.124	116	13,456	10.770
17	289	4.123	67	4,489	8.185	117	13,689	10.817
18	324	4.243	68	4,624	8.246	118	13,924	10.863
19	361	4.359	69	4,761	8.307	119	14,161	10.909
20	400	4.472	70	4,900	8.367	120	14,400	10.954
21	441	4.583	71	5,041	8.426	121	14,641	11.000
22	484	4.690	72	5,184	8.485	122	14,884	11.045
23	529	4.796	73	5,329	8.544	123	15,129	11.091
24	576	4.899	74	5,476	8.602	124	15,376	11.136
25	625	5.000	75	5,625	8.660	125	15,625	11.180
26	676	5.099	76	5,776	8.718	126	15,876	11.225
27	729	5.196	77	5,929	8.775	127	16,129	11.269
28	784	5.292	78	6,084	8.832	128	16,384	11.314
29	841	5.385	79	6,241	8.888	129	16,641	11.358
30	900	5.477	80	6,400	8.944	130	16,900	11.402
31	961	5.568	81	6,561	9.000	131	17,161	11.446
32	1,024	5.657	82	6,724	9.055	132	17,424	11.489
33	1,089	5.745	83	6,889	9.110	133	17,689	11.533
34	1,156	5.831	84	7,056	9.165	134	17,956	11.576
35	1,225	5.916	85	7,225	9.220	135	18,225	11.619
36	1,296	6.000	86	7,396	9.274	136	18,496	11.662
37	1,369	6.083	87	7,569	9.327	137	18,769	11.705
38	1,444	6.164	88	7,744	9.381	138	19,044	11.747
39	1,521	6.245	89	7,921	9.434	139	19,321	11.790
40	1,600	6.325	90	8,100	9.487	140	19,600	11.832
41	1,681	6.403	91	8,281	9.539	141	19,881	11.874
42	1,764	6.481	92	8,464	9.592	142	20,164	11.916
43	1,849	6.557	93	8,649	9.644	143	20,449	11.958
44	1,936	6.633	94	8,836	9.695	144	20,736	12.000
45	2,025	6.708	95	9,025	9.747	145	21,025	12.042
46	2,116	6.782	96	9,216	9.798	146	21,316	12.083
47	2,209	6.856	97	9,409	9.849	147	21,609	12.124
48	2,304	6.928	98	9,604	9.899	148	21,904	12.166
49	2,401	7.000	99	9,801	9.950	149	22,201	12.207
50	2,500	7.071	100	10,000	10.000	150	22,500	12.247

Fraction	Decimal	Fraction	Decimal
$\frac{1}{2}$	0.5	$\frac{5}{10}$	0.5
$\frac{1}{3}$	$0.\overline{3}$	$\frac{6}{10}$	0.6
$\frac{2}{3}$	$0.\overline{6}$	$\frac{7}{10}$	0.7
$\frac{1}{4}$	0.25	$\frac{8}{10}$	0.8
$\frac{2}{4}$	0.5	$\frac{9}{10}$	0.9
$\frac{3}{4}$	0.75	$\frac{1}{16}$	0.0625
$\frac{1}{5}$	0.2	$\frac{2}{16}$	0.125
$\frac{2}{5}$	0.4	$\frac{3}{16}$	0.1875
$\frac{3}{5}$	0.6	$\frac{4}{16}$	0.25
$\frac{4}{5}$	0.8	$\frac{5}{16}$	0.3125
$\frac{1}{8}$	0.125	$\frac{6}{16}$	0.375
$\frac{2}{8}$	0.25	$\frac{7}{16}$	0.4375
$\frac{3}{8}$	0.375	$\frac{8}{16}$	0.5
$\frac{4}{8}$	0.5	$\frac{9}{16}$	0.5625
$\frac{5}{8}$	0.625	$\frac{10}{16}$	0.625
$\frac{6}{8}$	0.75	$\frac{11}{16}$	0.6875
$\frac{7}{8}$	0.875	$\frac{12}{16}$	0.75
$\frac{1}{10}$	0.1	$\frac{13}{16}$	0.8125
$\frac{2}{10}$	0.2	$\frac{14}{16}$	0.875
$\frac{3}{10}$	0.3	$\frac{15}{16}$	0.9375
$\frac{4}{10}$	0.4		

Made in United States
Troutdale, OR
03/08/2024

18304328R00151